THE LOGICS OF PREFERENCE

EPISTEME

A SERIES IN THE FOUNDATIONAL,
METHODOLOGICAL, PHILOSOPHICAL, PSYCHOLOGICAL,
SOCIOLOGICAL, AND POLITICAL ASPECTS
OF THE SCIENCES, PURE AND APPLIED

Editor: MARIO BUNGE

Foundations and Philosophy of Science Unit, McGill University

Advisory Editorial Board:

RUTHERFORD ARIS, Chemistry, *University of Minnesota*
HUBERT M. BLALOCK, Sociology, *University of Washington*
GEORGE BUGLIARELLO, Engineering, *Polytechnic Institute of New York*
NOAM CHOMSKY, Linguistics, *MIT*
KARL W. DEUTSCH, Political Science, *Harvard University*
BRUNO FRITSCH, Economics, *E.T.H. Zürich*
ERWIN HIEBERT, History of Science, *Harvard University*
ARISTID LINDENMAYER, Biology, *University of Utrecht*
JOHN MYHILL, Mathematics, *SUNY at Buffalo*
JOHN MAYNARD SMITH, Biology, *University of Sussex*
RAIMO TUOMELA, Philosophy, *University of Helsinki*

VOLUME 14

NICHOLAS J. MOUTAFAKIS
Cleveland State University

THE LOGICS OF PREFERENCE

A Study of Prohairetic Logics in Twentieth Century Philosophy

D. REIDEL PUBLISHING COMPANY

A MEMBER OF THE KLUWER ACADEMIC PUBLISHERS GROUP

DORDRECHT / BOSTON / LANCASTER / TOKYO

Library of Congress Cataloging in Publication Data

Moutafakis, Nicholas J., 1941–
  The logics of preference: a study of prohairetic logics in twentieth century philosophy / by Nicholas J. Moutafakis.
    p. cm—(Episteme: v. 14)
  Bibliography: p.
  ISBN 90-277-2591-8
  1. Deontic logic. 2. Philosophy, Modern–20th century. I. Title.
II. Series: Episteme (D. Reidel); v. 14.
BC145.M68 1987
121′.8–dc 19   87-19710
                                                              CIP

Published by D. Reidel Publishing Company,
P.O. Box 17, 3300 AA Dordrecht, Holland.

Sold and distributed in the U.S.A. and Canada
by Kluwer Academic Publishers,
101 Philip Drive, Assinippi Park, Norwell, MA 02061, U.S.A.

In all other countries, sold and distributed
by Kluwer Academic Publishers Group,
P.O. Box 322, 3300 AH Dordrecht, Holland.

All Rights Reserved
© 1987 by D. Reidel Publishing Company, Dordrecht, Holland
No part of the material protected by this copyright notice may be reproduced or
utilized in any form or by any means, electronic or mechanical
including photocopying, recording or by any information storage and
retrieval system, without written permission from the copyright owner

Printed in The Netherlands

Richard M. Martin
"In memoriam"

".... 'ω 'άνδρεζ, δάνατον εκφυγείν, 'αλλὰ πολυ χαλεπώτερον πονηρίαν, θᾶττον γὰρ θανάτου θεῖ, ..."

ΑΠΟΛΟΓΙΑ ΣΩΚΡΑΤΟΥΣ...29,39E

TABLE OF CONTENTS

Introduction 1

PART ONE

The Objectivist Approach Toward the Formalization of Preferences

Chapter 1.   Prototheoretic Attempts Toward a Logic of Preference 13
Chapter 2.   Aristotelean Reflections in Richard M. Martin's
             Extensionalized Pragmatics of Preference 57
Chapter 3.   Rescher's Logic of Preference and Linguistic Analysis 89
Chapter 4.   Richard C. Jeffrey's Logic of First and Higher-Order
             Preferences 119

PART TWO

The Subjectivist Approach Toward the Formalization of Preferences

Chapter 5.   Soren Hallden's "Puristic" Logic of the Better and Same 167
Chapter 6.   The Many Modal Interpretations of Prohairetic Logic:
             Aqvist, Chisholm, Sosa and Hansson 191
Chapter 7.   Von Wright's Logic of
             Propositions Expressing Preferences 221
Chapter 8.   Hochberg on the Logic of
             "Extrinsic Epistemic Preferability" 257

Postscript 269
Selected Bibliography 279
Name Index 285
Subject Index 287

Introduction

With characteristic incisiveness Georg Henrik von Wright identifies *prohairetic* logic (i.e. the logic of preference) as the core of a general theory of value concepts. Essentially, this nucleus involves *the logical study of acts from the point of view of their preferability.*[1] (italics added) Though the term *prohairesis* is found in Plato, as well as in Aristotle's treatment of the relations of preference, it is von Wright who introduces this word into contemporary analytical philosophy, and succinctly specifies the philosophical dimensions it encompasses.

The above emphasis upon the *philosophical* study of the formalization of preferences is a matter of utmost importance for understanding the type of inquiry this investigation attempts to initiate. Over the past one hundred years the literature on general theories of subjective utility has become massive, where one considers the work done in psychometrics, econometrics, statistical theories, probability theories, etc., etc. Histories in these areas are strong in tracing various evolutions in the development of the concept of preference in decision-making. However, what has not been investigated with sustained attention are the fundamentally philosophical inquiries into the formalization of preference-relations.

In surveying the purely philosophical explorations in this area, it is often impossible to separate the purported philosophical implications of these inquiries from the relevance this work has in other areas. For example, Ramsey's contribution is often seen as a stunning innovation in the area of subjective utility theory. As such, it is usually taken as a reaction to the at the time popular Keynesian rendition of probability theory.

Yet this is a much too restricted view on what Ramsey had to say from a philosophical standpoint. The latter's views on partial belief and risk are introduced into the philosophical literature by Davidson, who recognizes their importance as a basis for further inquiry into the logic of preference. Thus an exegesis of Ramsey's contribution, as it will be undertaken in the opening chapter of this book, demands an in-depth account of what had occurred in

---

[1] Von Wright, Georg Henrik "The Logic of Action - A Sketch," in *The Logic of Decision and Action*, University of Pittsburgh Press, 1967, pp. 135-136.

the evolution of the concept of preference in the general study of decision theory. A delicate balance must be struck therefore between what is needed for purposes of explication from other disciplines, while retaining the primary emphasis upon the philosophical import of this and other attempts, as they evolve in this unique body of literature.

All of this indicates that the aim of the investigation which follows is not to account for a general theory of subjective cardinal utility, but to present a detailed discussion of the fundamentally philosophical inquiries into the formalization of preference-relations, as the *core* of a theory of valuation.

A progressively more accented concentration upon the manner by which preferences are conveyed in ordinary discourse becomes the single most distinctive feature in the analyses to be discussed, and serves to sharply demarcate the philosophical interest in the formalization of preferences from the analysis of preference in other areas, e.g., decision theory and econometrics. In every philosophical investigation to be considered, one finds varying levels of concern expressed as to how well their attempts capture the "sense" of preference-relations as manifested in some linguistic context, whether that of everyday discourse or of some more circumspect sort. Indeed, one may argue that this difference underlies Nicholas Rescher's broad differentiation between the "logico-philosophical" attempts at a logic of preference, and the "mathematico-economic" ones.

Whereas the cleavage between the philosophical interest in the concept of preference is readily discernible in contrast to how preference is handled in other disciplines, the account of how logics of preference differ amongst each other within philosophy itself demands that attention be directed upon details of approach in every case. Surely, Ramsey's aim in formalizing preferences in "Truth and Probability" is very different from what von Wright attempted to accomplish in *The Logic of Preference*. Even in comparing Ramsey's work in conjunction with Richard C. Jeffrey's, where the latter is normally seen as a refinement of what the former had achieved, there is a noticeable difference between the two. For Ramsey's concern is with the analysis of partial belief under conditions of risk, whereas Jeffrey is concerned with presenting a formalization of deliberation.[2] These kinds of differences are supremely important where one endeavors to present a cogent account of the evolutionary phases in this particular area of philosophical inquiry.

---

[2] Vickers, John M., *Belief and Probability*, D. Reidel Publishing Company, Dordrecht-Holland/Boston-U.S.A., 1976, pp. 59-60.

So as to manage the wide diversity of opinion within this body of philosophical literature, two broad lines of inquiry will be identified as helpful in classifying the writings to be discussed. In the first grouping, beginning with Ramsey's work, one encounters a highly descriptive approach to the formalization of what one could do under varying conditions influencing belief and risk; couched in terms of some logico-mathematical model. Generally, this *"extensional approach"* is further manifested in Davidson's early work, as well as in the contribution provided by Herbert Bohnert. What changes in each case, however, is a refinement of the notion of accurate empirical measurement, as applied to the characterization of choice. Thus, Davidson cites the need to introduce an improvement in the methodology of numerical attribution, one which allows for "logical consistency", while also improving upon Ramsey's reliance upon the interval scale in determining valuation. By contrast, Bohnert's attempt concentrates upon disowning the use of frequency probability as employed in the von Neumann-Morgenstern analysis of choice, and favored by Davidson and his associates early on. Rather, Bohnert elected to pursue the formalization of preference expressions along the lines of Carnap's theory of logical probability, i.e., Carnap's *"probability$_1$"*. Moreover, these refinements are presented within a context of the overriding concern with linguistic usage mentioned earlier; which is in essence the interest in the degree of "fit" these descriptions have with respect to how preferences are conveyed in discourse about the "real world."

A similar analytical attitude is operating in the manner by which R.M. Martin endeavors to articulate the formalization of preference-relations within a specialized setting of testing and experimentation. Martin's investigation is perhaps even more sensitive than any other of its time to the nuance of discourse pertaining to the expression of preferences, specifically within the mode of an extensionalized pragmatics. His use of the basics of propositional logic to characterize the hierarchies such relations exhibit within this context leans heavily toward a dependence upon the sense of scientific discourse, which is more restrictive than that of natural language. In this respect, therefore, one may say of his attempt that it is less pervasive in intent than Ramsey's or Davidson's.

Nicholas Rescher's effort is also part of this grouping, because of his use of numerical attribution to describe the value of preferential states of affairs. Rescher operates on the assumption that one can devise a logic of preference which exhibits a fundamental mathematical rigor, one which can be said to underlie the necessity preference-relations convey in ordinary discourse. Apart from what is said on his success in this direction of inquiry, Rescher's work is unique in that he attempts to apply techniques developed in econometrics to a special area of philosophical analysis, i.e., that which pertains to the clarification of discourse involving choice.

Jeffrey is within this grouping as well, since he directs his work on the formalization of preference expressions as a progression beyond what Ramsey had accomplished. The former insists that both evaluative preference and cognitive preference must be handled within a "unified" mathematical theory, one which has propositions as its objects, to which both subjective probabilities and desirabilities could be ascribed.

The second grouping is not radically different from the first, with respect to the fact that it also contains work which exhibits an interest in the logical analysis of expressions of preference. What does distinguish it from the attempts of authors already mentioned is that here preference is seen in terms of its *intrinsic* meaning, i.e., in the sense where one prefers something in and of itself, under conditions of *ceteris paribus*. Noticeably absent in this collection of writings is any consideration regarding extensional factors, such as risk, and the attempted application of probability theory. Along these lines one also finds a pronounced concern with the evaluative aspect of preference with reference to ordinary discourse made only as a way of noting the comparative strength in the expressive power of the logics which are evolved. This constitutes a further idiosyncrasy which is characteristic of this latter grouping, namely the allusion to ordinary discourse is made so as to underscore the advantage of the logics which emerge, rather than to show that these logics are in some way to be taken as embodied in ordinary discourse, or as descriptive of it in some sense.

First in the second grouping is Soren Hallden who in *The Logic of 'Better'* employs propositional logic to formalize a "puristic" logic of better, one which becomes in its own right a model for all subsequent logics of preference, dealing with the *intrinsically* preferable. Authors included in this grouping are acutely sensitive to the goal Hallden is the first to envision, namely to develop a logic which is unencumbered by the vagaries often encountered where an evaluative term such as 'better' is conveyed within ordinary language.

Extrapolating from Hallden's approach is the formalized account of preference expressions proposed by Lennart Aqvist. The latter's innovation is to introduce a calculus of preference which is an extension of a deontic logic. Aqvist's allusion to the obligatory, the forbidden, and the value-indifferent, as a means of securing the formal properties of preference-relations in the intentional sense of the term 'preference', introduces a uniquely moral emphasis into this type of inquiry. The application of deontic logic here is held to be absolutely necessary primarily because it avoids the inherent ambiguity ordinary discourse exhibits where preference is conveyed, the latter standing in the way of understanding the notion of preference with any degree of precision.

Concentration upon the moral connotations of the notion of preference is continued in the work by Roderick Chisholm and Ernest Sosa. These authors employ examples from moral discourse pertaining to a hedonistic conception of the intrinsically preferable as foils from which to project a general logic of preference principles. In the course of their investigation they become acutely aware of a paradox which their very mode of inquiry incurs. For propositional logic presupposes the formalization of propositions as sententially expressed meanings, whereas in their view the logic of preference must deal exclusively with states of affairs, having a much different ontological status.

One has here the early realization of a difficulty which is not unlike the problem perceived by writers comprising the first grouping, going as far back as F.P. Ramsey. Basically, it is the question of whether a logic of preference is properly said to be a logic about things, i.e., it is a logic *de re*, or whether it is a logic whose object is linguistic expressions about what is preferred, i.e., it is a logic *de dicto*. Chisholm and Sosa resign themselves to their logic having this internal ambiguity, for they see no way of resolving the *de dicto/de re* opposition.

Benght Hansson is equally aware of the ambiguity diverse contexts promulgate upon the formalization of a preference logic, especially where one tries to articulate such a logic from the foundation of some single discipline, such as economics or jurisprudence. He thus attempts to present a logic which totally transcends identification with any possible context, even where that context is that of propositional logic itself. In Hansson's view the logic of preference must be of a sufficiently diaphanous character so that it will have the flexibility to express the rich variety preference-relations are found to have in countless numbers of disciplines. In the final analysis, he believes he has achieved a logic which is altogether independent of the confining restraints poised by the very basic differentiation between intrinsity and extrinsity, a distinction which is essential to the entire field discussed by this book.

In contrast to Hansson's search for the best possible logic of preference one has Georg Henrik von Wright's towering contribution. His activity in the field extends over a period of twenty-five years, covering both critical essays and his own treatise, *The Logic of Preference*. Von Wright was among the first to insist that philosophically the logic of preference must exhibit its own unique qualitative character, and thus cannot be sought with an extensional mode of analysis, where much is left to quantitative factors such as risk, utility, and subjective probability. Moreover, its independence must also be seen apart from any deontological analyses, since the latter cannot do justice to the complex anthropological and axiological facets presupposed in the notion of preference. On the other hand, this logic should not be of such a nature that one wonders what its subject matter is all about, as is the case with what Hansson presents.

While pursuing his analysis of intrinsic preferences, under conditions of *ceteris paribus*, von Wright deals with the issue first raised by Ramsey, namely what the implications are when the *de dicto/de re* distinction is introduced into philosophical discussions regarding preferences. Von Wright's way of handling the problem is to conceive the objects of preference as "proposition-like" entities, which is a way of having them be "thing-like," while also not divesting them of a linguistic status. This peculiar ontology is taken by von Wright as providing a justification for the application of propositional logic, while preserving the appearance that preferences in the resulting logic deal with concrete events and states of affairs.

Finally, the work of Gary Hochberg is presented as an example of an interesting thrust in the direction of trying to find a common ground for the two groupings discussed above. His analysis is unconsciously reflective of von Wright's crucial insight that *prohairetic* logic is at the core of a theory of valuation. Hochberg's strategy is to define certain key terms pertaining to the *intrinsically* preferable, such as "indifference," "sameness," "goodness," etc., and then to offer a complementary series of definitions concerning terms having to do with the *extrinsically* preferable, e.g., reasonable, acceptable, favorable, etc. Hochberg then proceeds to present a logic for the latter, i.e. preference in an extrinsic sense, through a consideration of conditions covering epistemic relevance within a context of preference. The core of the resulting logic, however, contains the nucleus of the definitions that relate to the *intrinsically* preferable terms with which he began his inquiry.

Justifiably, one may ask whether the authors alluded to above comprise a complete account of all who have worked in this area. Surely the answer here must be *no*, since one finds commentaries on the concept of preference in Aristotle, Brentano, Meinong, Kenneth Arrow, De Finetti, and a host of others. The problem in the presentation of these latter contributions is that in themselves they do not fit in an orderly historical sequence, where one aims to concentrate solely upon the evolution of the logic of preference in the present century. On the other hand, no study which makes any pretense to thoroughness can afford to ignore the important work these additional authors performed. Consequently Aristotle's pioneering insights on the concept of preference are included here by discussing how uncannily close he comes to a conception of preference found in R.M. Martin's extensionalized pragmatics. Moreover, Brentano and Meinong are discussed as a means of giving depth to Roderick M. Chisholm's and Ernest Sosa's phenomenological analysis of the concept of preference. Indeed, the latter allude to Brentano and Meinong as not only important contributors to this field, but they use the approach these authors develop as a foundation for their own logic of preference. Kenneth Arrow's views on certain aspects of a general theory of choice are introduced in the same fashion. First they are discussed in connection with a criticism of Ramsey's theory of subjective probability. Secondly, they are considered in concert with Davidson's allusion to the von Neumann-Morgenstern theorem while presenting his own logic of rational choice. Surely Arrow's work deserves a chapter of its own, were this a general study in the area of the evolution of the concept of subjective cardinal utility. However, to have gone into such a direction would have involved the danger of being led astray by economic theories pertaining to pricing and decision-making. De Finetti's work is handled with similar caution. The latter's views are taken up as they relate to aspects of the criticism of Richard C. Jeffrey's assumptions concerning the function of a logic of preference, where probability theory is involved. To have gone deeper into De Finetti, especially with respect to his position on the interchangeability of events, may have led the discussion far afield.

Ultimately, one must face the serious question as to why a book on the formalization of preferences. Apart from the pressing editorial need to present in some cogent way the wide scattering of material in journals and monographs ranging over a period of fifty years, there is the philosophical importance of such a study, especially as it pertains to the construction of theories concerning value.

Recalling von Wright's perceptive remarks regarding the centrality of a logic of preference, one wonders whether meta-ethicists would not progress further in explaining the logic of moral discourse were they to understand the complexity of issues encountered by those who have endeavored to achieve a formalization of the elusive preference-relation. Surely the intersection of axiological and praxiological considerations, being constitutive of the concept of preference under any conceivable interpretation, is central to moral theorizing in general, and cannot be dismissed as inconsequential or peripheral.

Moreover, the issues raised above on the applicability of formal logic in expressing the formal sense of preference in everyday discourse involving values are key issues, in that they are preparatory for dealing with problems about the implications among moral choices. Given the interconnected character of human society, preference cannot be considered as totally isolated, so that there is no effect between the choice of one person and that of some other. A workable logic of preference can serve as a guide toward clarifying the implicative character of the relation of preference, considered in the context of meta-ethical inquiry.

The importance of this study, however, goes beyond that of simply being able to provide a clarificatory tool for meta-systematic inquiries as to the nature of discourse involving choice. There are also very practical uses for logics of preference, and these should not be lost sight of when discussing the above. The specificity of this inquiry is not therefor limited by the fact it has uses at the highest levels of philosophical investigation.

One immediate point of tangence is the evident importance of having a logic of preference as a foundation for a theory of valuation, reflective of instrumentalism in the manner suggested by John Dewey. Surely where human action is seen extensionally as "an ongoing human activity," then value within such a framework is taken as nonstatic, that is, as valuation *for* some purpose or end. For this reason a theory of preference is ideally suited for the prioritizing of one's actions toward the realization of some clearly specified end. In essence, where valuation is seen in terms of appraisal or *estimate*, rather than in terms of *esteem*, then a theory of preference has significance as a logical foundation of the language for a general theory of value. This at least appears to be a most productive function for theories of preference, where one considers value in entirely extensional terms. Philosophers, however, when working with theories of instrumental value seem oblivious to literature on the logic of preference, much to the detriment of the views they endeavor to present.

Even in cases where the approach to the concept of value is that of the prizing or *esteem* of some end in and of itself, independently of any consideration of the means, one may find some use in adopting certain conclusions reached on the formalization of preferences. Surely where the intrinsic goodness of an end becomes the fundamental conceptual basis of a logic of preference, as is found to be the case with Hallden, Aqvist and von Wright, then analysts choosing a moral objectivist stance can discover a useful way of structuring their perceptions of value by alluding to this literature. However, here one finds a similar inattentiveness to the research which has been done in *prohairetic* logics as pointed out above.

Moreover, and apart from Mullen's aphoristic remarks pertaining to the feasibility of any logic of preference, one can now see that the utility of such logics ultimately depends upon the philosophical methodology one elects to work with.[3] Consequently, unless the broader philosophical question is considered first, i.e. whether the formalization of preferences is properly conceived either in an intentional or an extensional mode, or perhaps as even a hybrid of these two, as suggested by G. Hochberg, it is premature to dismiss the philosophical importance of these logics, though granted none has as yet emerged as being beyond reproach.

Over the period of time required for the preparation of this manuscript I have had occasion to confer with a number of scholars who in the majority of cases have contributed significantly to the study of the development of prohairetic logics. To these I would like to express my gratitude for their encouragement in supporting my inquiries in this area. Their counsel on the manner by which I have critiqued their writings has been most helpful. In alphabetical order these are: Lennart Aqvist, Herbert G. Bohnert, Roderick M. Chisholm, Donald Davidson, Peter C. Fishburn, Soren Hallden, Haig Hatchadorian, Gary Hochberg, Richard C. Jeffrey, Richard M. Martin, Nicholas Rescher, Patrick Suppes, Georg Henrik von Wright and Alan White. Through the enjoyable course of our discussions and correspondence variances emerged as to how one properly interprets the various views on the formalization of preference expressions. Yet agreement was uniform on the importance of such a work both philosophically, and as a way of filling the need for organizing the literature on this issue.

Finally, I would like to thank Professor Richard M. Fox for providing me with the moral support this project required through its long and arduous turns. Also I am deeply indebted to the fine secretarial help I received from Mrs. Cynthia Bellinger, Mrs. Mary Persanyi, and Mrs. Margaret Sholtis, without whose assistance this manuscript would have been incomplete.

My special gratitude to Professor A. Harry Andrist and the Office of Printing and Composition Services, Cleveland State University.

---

[3] Mullen, John D., "Does the Logic of Preference Rest on a Mistake," *Metaphilosophy*, 10, (1979).

PART ONE

The Objectivist Approach Toward the Formalization of Preferences

Chapter 1.

Prototheoretic Attempts Toward a Logic of Preference

Contemporary attempts in arriving at logics of preference, within a strictly philosophical context, begin formally with Frank P. Ramsey's effort, published posthumously in 1931. Ramsey's concern was to find a means by which to express mathematically "uncertain partial belief". This is belief operative in conditions under which a choice of options is available to the subject, none of which is he fully certain will satisfy securing a particular objective. Ramsey's work was revolutionary, not because of its success in realizing what it set out to do, but because of Ramsey's conviction that he could discover a means by which to objectify the "mental state" of uncertain belief, i.e., to extensionalize it in terms which will allow for the mathematical axiomatization of choice. Recent literature on the logic of preference does not record any earlier philosophical effort, in which such a specific interest is illustrated in the area of Prohairetic Logic, that is the logic of rational action and choice.

As with Hume's *A Treatise of Human Nature*, Ramsey's endeavor in *Truth and Probability* to initiate discussion in a new area "... fell still-born from the press. ..." Active inquiry within the analytical tradition had to wait almost a quarter of a century, until attention was directed again to the logic of preference, in works by Davidson, McKinsey, Suppes and Bohnert. One can only surmise the reasons for this absence of discussion from Ramsey's time to that of the writers mentioned. Perhaps the dominance of Logical Positivism and its insistence upon sharp dichotomies between cognitive and emotive signification suggested to investigators that there was no way of verifying claims dealing with dispositional attitudes, such as those which partial belief apparently involves, without having to utter statements empirically not verifiable. Though such conjecture cannot point precisely to the actual reason(s) influencing a subject's de-emphasis within ongoing philosophical discussion, at the very least it can suggest the kind of intellectual climate which may have thwarted pursuit of this vital area of philosophical investigation.

By the mid 50's, work re-appears within philosophical journals, acknowledging the seminal beginning which Ramsey made in this area. However, explorations such as the ones initiated by Davidson and his associates approach the issue of the axiomatization of preference from the viewpoint of what economists, in particular von Neumann and Morgenstern, had done in setting forth utility theories governing the rationalization of choice. Davidson's allusion to such theories was primarily the result of a scarcity of analytical tools within any philosophical tradition, which would have enabled investigators to handle the subtle difficulties which preference seemed to involve.

Furthermore, adaptation of an economico-mathematical analysis precipitates something of a shift from Ramsey. For instead of considering "uncertain partial belief" as a psychological state, interpretable in terms of modes of action one may or may not take in a given situation, one now begins with a constructed conceptual framework of agent and environment, articulated on the maxim that rational choice is definable only in terms of logical necessity, dictated in a theory of frequency probabilities. Consequently, in Davidson's effort, unlike that of Ramsey's, one finds a greater emphasis placed upon the logical consistency of preference rankings, at the expense of their possible idiosyncratic nature, as manifested in ordinary discourse. With the subsequent use of analyses found in economics, one has introduced in the philosophical investigation of preference a richer interval scale. This constitutes a further departure from Ramsey's approach with its emphasis and reliance on a simple ratio scale of measurement.

Yet, even where analyses by economists are adopted for the purpose of clarifying philosophical discussions, the reservation most often voiced is that of the nature of the relation between theoretical construction and the reality it is intended to describe. Though Davidson alludes to the flexibility which the probability theory provides in describing formally the weighing of alternatives, one finds him also recognizing how far short these theories fall in expressing the complexity of "lived experience," as it is involved in expressions of preference. Remarkably, the unease Ramsey reveals concerning his ability to express logically the subjective probabilities of uncertain belief continues with Davidson, McKinsey and Suppes, where they endeavor to apply more technical means to express the determination of comparative success among alternative modes of action.

Generally, one can only hint at the treasury of philosophical concerns which a study of preference involves, e.g.: (1) the interplay of issues dealing with the correct psychological explanation of the mental state of expectation which preference seems to suggest; (2) the power in logical accounts for expressing behavioral action; (3) the part value judgments may or may not play in the manifestation of hierarchies of preference, etc. — all these are touched upon by such inquiry. This is seen at the very beginning, with the question of whether preference must reflect rational choice. One observes Ramsey emphatically maintaining that "rational choice" must not fail to be characterized in a way which takes into account the psychological element, nor should logical requirements dominate the role of desire. Yet, in contrast, Davidson emphasizes logical consistency as the sole basis of "rational choice." As will be seen, the concerns often voiced by these investigators over the adequacy of their analyses bring into focus the complexity of solutions which (1) try to give an objective account of a subjective phenomenon; (2) by-pass the role of value judgments which may serve to upset the sacrosanct "logicality" of preference, and (3) recognize the distance which apparently will always persist between theories and the realities they are intended to picture.

However, the difficulties experienced above stem precisely from a failure to concentrate upon the meaning of "preference" in ordinary discourse. This is to say that preference is not analyzed in terms of its linguistic function as a "form of life." Rather, Ramsey proceeds in his analysis from the viewpoint of an early Wittgensteinian, where language is fractured into atomic propositions. On the other hand, Davidson, McKinsey and Suppes proceed by completely avoiding any discussion which may involve them in the "... inadequate and inflexible resources of ordinary language. ..."[1] The price which both they and Ramsey must pay for their inattentiveness to the linguistic element is a constant reiteration of doubt on whether their formal analyses are fully reflective of the subtleties of preference, as these emerge in local contexts. This repeated expression of misgiving is particularly ironic in the work of Davidson and his associates, because of their expressed wish to offer a formalization of preference which departs from the inflexible resources of ordinary language.

The ensuing discussion attempts to place in historical perspective the first formal inquiries into the development of calculi for preference, as found in the writing of F.P. Ramsey, and in the collective effort by Davidson, McKinsey and Suppes. The object is to show how both of these important attempts founder on the shoals of linguistic use. To date, no attention has been given to the historical evolution of these theories or to their deficiencies from the viewpoint of linguistic analysis.

---

[1] Davidson, McKinsey and Suppes, "Outlines of a Formal Theory of Value, I", *Philosophy of Science*, Vol. 22, 1955, p. 142.

I. F.P. Ramsey on the Analysis of Degrees of Belief

It is not in exaggeration that Georg Henrik von Wright refers to Ramsey's essay, *Truth and Probability*, as "ingenious."[2] Without doubt, Ramsey's insights into the logic of preference have not been surpassed since he first expressed them in 1931. If any of his numerous achievements will last as a living memorial to his brief life, this work on the analysis of degrees of belief will long endure as a tribute to his brilliant career.

Ramsey approaches the study of preference from the perspective of how one is to devise a means of measuring degrees of belief. His discussion emerges as a consequence of his criticism of Keynes' view concerning the kind of evidence which governs the determination of belief in the conclusion of inductive arguments. To sketch briefly the stage on which Ramsey presents his analysis: he argues against the idea that the degree of belief one attaches to the conclusion of an inductive argument is directly determined by the degree of belief one attaches to the premises of that argument. He reasons that one cannot treat the degree of belief in an inductive generalization merely as an accumulative factor, stemming from one's belief in its premises. Rather, measurement of one's degree of belief in the conclusion of inductive argumentation must also take into account the meaning of that conclusion, in relation to the context of an individual who is faced with the prospect of accepting the consequences of acting upon that conclusion.[3]

The point Ramsey is making is that the degree of belief in the conclusion of inductive arguments is a matter separate from the determination of the truth of such conclusions. Ordinarily, one would say that the truth of an inductive generalization is directly dependent upon the number of its confirming instances. However, one's belief in the conclusion of such arguments is influenced by what the conclusion allows or does not allow the subject to do in a particular context. One's degree of belief in such a conclusion is influenced by praxiological considerations. Hence, the mechanics of why a conclusion to an inductive argument is more or less probable do not constitute the reasons for determining one's degree of belief in that conclusion. For example, the high probability of the generalization "All metals expand when heated" results from the overwhelming number of cases which confirm its agreement with the facts. However, one's degree of belief in this generalization is comprehended

---

[2] Von Wright, Georg Henrik, *The Logic of Preference*, Edingburgh University Press, 1963, p. 17.

[3] Ramsey, F. P., *The Foundations of Mathematics and Other Logical Essays*, London: Kegan Paul, 1931, pp. 160-165.

in terms of the instrumental value of this generalization for the subject. In other words, the subject believes that the conclusion, as true, will enable him to achieve some end, for which the conclusion is but a justification. Thus, the degree of belief is a function of the pragmatic context of how the subject is willing to act on the truth of an expression of fact. In general terms, the issue of the strength of an inductive conclusion is an epistemological matter, while the question of whether to act upon the information conveyed by that conclusion also involves praxiological considerations. Hence, Ramsey is bringing out an important and hitherto unrealized distinction between those factors which go into confirming the truth of inductive generalizations, and those factors which go into determining one's degree of belief in the conclusions of inductive arguments. Essentially, the distinction comes down to (1) considering the extent of supporting evidence in inductive reasoning, and (2) considering the extent of supporting evidence in ones acting upon the result of an inductive generalization. It is these fine points which Keynes had not seen, and which Ramsey is quick to point out.

The innovative importance of Ramsey's analysis of partial belief is greater where one takes into account its value in contrast to Thomas Bayes' work, published in 1763. Though reference to Bayes will occur again in connection with Richard C. Jeffrey's analysis of the preference-relation within a context of probability theory, Bayes also merits mention here for two reasons. First, Ramsey was aware of Bayesian analysis, and saw his own work as an improvement over it. Secondly, Bayes defined probability as the ratio between the value of the expectation for the occurrence of some event or proposition being true, and the value of that occurrence or proposition itself. Surely, where Bayes speaks of "probability" one can insert the words "strength of belief in," and since his work concerns itself with the idea of "degree of belief," it relates to Ramsey's investigation in a very important way.

Bayes assumed that his numerical ratio was invariant and uniformly applicable to both expectations and states of affairs. Nonetheless, cases can be presented where a numerical or monetary scale of measurement breaks down when it comes to applying the above ratio. John Vickers identifies some of these unyielding examples in his recent book *Belief and Probability*. One instance is where it is conceivable that an individual may be willing to insure against a $3,000 loss with a yearly premium of $100.00, whereas he would only be willing to pay a $40.00 premium against a $1,500 loss. The consistency of Bayes' ratio is at stake here, though the difference in premium amounts is perfectly concordant with a person's tolerance of monetary loss, given his personal and therefore idiosyncratic circumstances. The drawback of Bayes' account of "strength of belief" or probability is in its clashing with the real world, where the attribution of numerical value to events and propositions is not as uncontroversial as Bayes evidently supposed. Basically, there is no determinable natural zero value point when it comes to the valuation of states of affairs, as there is for degrees of temperature or weight. Consequently, there cannot be an invariance in the relation between expectations and actual outcome, contrary to Bayes' assumption in presenting his probability ratio.[4]

As will be seen, Ramsey's solution to the above difficulty will be to concentrate upon that aspect of the partial belief ratio where invariance is guaranteed. This is found when the situation is defined in such a way that an individual considers the risk of his betting on an option given that he cannot lose everything should that alternative not come forth, while also assuming that he is indifferent to whatever the final outcome is. The details of Ramsey's position will be presented further on. Here it is sufficient to point to how he is attempting to present a rigorous account of the idea of degree of belief, one which is more sophisticated in the sense that it does not bind the believer into an all or nothing situation, where the agent must accept consequences which are easily controverted by the variety and complexity of everyday circumstances. By removing the agent's personal interest in the choice he is making in placing the bet, Ramsey is showing how an environment can be defined which allows for the play of strictly mathematical relations. This in turn serves to provide a formalization of partial belief which is more flexible in its tangence with the actual world of things. The success of what Ramsey proposes must be assessed further on, where questions can be raised as to the plausibility of having bets as part of the analysis of degrees of belief. Nonetheless, at this early stage it is clear that he sees the need to present a formal analysis of one's acceptance of a proposition in a way which is as context neutral as possible. Insofar as he succeeds in this endeavor to a certain extent, he certainly has improved upon Bayes' perception of the idea of the strength of a belief.

By way of setting up a context for his own presentation, Ramsey recognizes the need to adopt some psychological stance on the subjective nature of partial belief, so as to improve its access to philosophical investigation. He does not avoid considering partial belief as a dispositional attitude, but rather develops a technique which minimizes the difficulty when dealing with such essentially private phenomena. By attempting to find a means of extensionalizing the factors for adopting one course of action rather than another, Ram-

---

[4] Vickers, John, *Belief and Probability*, pp.46-48.

sey seeks to overcome the supposed inaccessibility of such experience. Admittedly, though he realizes that the theory he elects to adopt is "universally discarded", he views it as a very important heuristic device. Thus, from Ramsey's pioneering perspective, it is nonetheless a theory which can play a fruitful role in explaining the limited domain of the phenomena which his discussion covers. Specifically, it is the view that one acts in a fashion which will most readily secure one's object of desire. It is important to note that Ramsey merely uses this as a heuristic device to launch his analysis, which is articulated in an essentially behavioristic context. Thus, he should not be taken as attempting to justify this theory in terms of the logical analysis which follows. Rather, what he says concerning the measurable determination of belief can be taken separately from the veracity of the psychological background which helps initiate his investigation.

Ramsey's assumption of a psychological context of action so as to achieve the most directly rational means to a goal reveals the basic difference between what he is trying to accomplish, and what subsequent attempts in psychometrics aimed to realize when analyzing choice. It should be noted that Ramsey was interested in the presentation of a theory of *consistent* subjective probability, whereas for the most part, psychometricians are concerned with the notion of preference, based upon the measurement of actual utility. As will be shown in connection with Davidson's contribution, it is necessary for those doing psychometrics to realize that the relative correlations they observe in their studies of human behavior could be more cogently explained if a framework such as Ramsey's was adopted, where the consistency of choice is held as foremost in importance.

The unease psychologists feel toward the introduction of logical theories to finesse their accounts of human behavior pertaining to preference is apparent in an excerpt from R.D.Luce and P.Suppes' *Preference, Utility and Subjective Probability*, (p.253):

> "Our attitude in this chapter is primarily psychological: we ask whether a theory seems to describe behavior, not whether it characterizes a rational man; we report experiments, although admittedly not enough experimental work has been done to provide us with either completely satisfactory designs or highly reliable phenomena; and we explore some of the relations between theories of preference and other psychological theories. At the same time, we try to recount the normative considerations that led originally to the theories and to cite the more important rationalistic criticisms ... leveled against them."

Assuming the existence of a subject who has no doubts about anything but does have certain opinions about all propositions, one can claim within the parameters of the given psychological framework that he will invariably choose the course of action which, in his view, will lead to the greatest amount of good. Ramsey points out that "good" here is not to be taken in an ethical sense, but in the sense of one's feeling of aversion to or desire for the object. Thus, by employing what Ramsey calls the "psychological law" that a person acts so as to achieve the greatest numerical expectation, one is soon in the position of having a means of measuring the subject's degree of belief in a given proposition. Quite simply, one's degree of belief in proposition p would be expressed as the action the subject would perform toward the realization of the good or the evils which result from acting upon the truth of p, over a period of time. Formally, one's degree of belief in p is $\frac{m}{n}$ where 'm' expresses the number of cases where his action makes p true (or false), and 'n' stands for the number of times attempted.[5]

It is vital to recall here that Ramsey is not seeking a definition of belief. Rather, he seeks to devise a means of measuring one's *degree* of belief. This he achieves by introducing the notion of preference as the numerically measurable and additive goal some agent desires. Thus "the degree of belief in p" is seen in terms of what the agent would choose to do over a period of time so as to achieve some end which makes p true. Hence, Ramsey is working with what von Wright will come to identify as the context of "extrinsic preference", which is preference considered in terms of means to ends.[6]

On many occasions Ramsey alludes to the work of Charles S. Peirce by way of justifying the conceptual divisions he proposes between deductive and inductive inference. It appears, however, that his indebtedness to Peirce extends beyond the central concern of determining the proper logical analysis of probability. His rendition of belief in terms of the actions one would perform is strongly reminiscent of Peirce's analysis of the evolution of belief within patterns of action. Surely Peirce's conception of the state of belief as action leading to cessation of doubt is reflected in Ramsey's example of how the ratio of one course of action as opposed to another comes to define one's degree of belief in a proposition concerning the utility of an action toward securing some goal.[7]

---

[5] Ramsey, F.P., *The Foundations of Mathematics*, pp. 172-174.

[6] Von Wright, Georg Hendrik, *The Logic of Preference*, p. 14.

[7] Peirce, Charles Sanders, "The Fixation of Belief,": *Philosophical Writings of Peirce*, edited by Justis Buchler, Dover Publications, Inc., New York, 1955, p. 11.

However, from a purely theoretical perspective, Ramsey articulates his formalization of preference as an extension of the logic of consistency, which is operative in the determination of the measurement of degrees of belief, and where beliefs are expressed as propositions concerning choices, i.e. preferences. The laws of preference he presents are thus applications of the laws of probability, which must be followed if the agent is to be conceived as capable of *intelligent* or "rational" choice.[8] Consequently, the axioms Ramsey sets forth and the theorems he demonstrates are purely formal. This is to say that they are axioms and theorems which attain their rigor solely on the basis of the logical necessity they are made to reflect.

When extending formal logic to explain consistent partial belief Ramsey finds that the relations uncovered reflect what follows necessarily by definition ('ανάγκη λέγειν) but not what follows necessarily by the objective state of things ('ανάγκη 'είναι). Thus "the calculus of objective partial belief cannot be numerically interpreted as a body of objective tautology." By contrast, purely formal logic has both 'ανάγκη λέγειν and 'ανάγκη 'είναι, in the sense that a necessarily true proposition is also true of the description of things in the formal language.[9] Since Ramsey's logic of preference is lodged within the context of hypothetical occasions, the agent is conceived as choosing so as to satisfy a theoretical degree of mathematical expectation. For this reason Ramsey states at the outset of the articulation of his system that his is only a *schematic* representation of real life.[10]

Ramsey's sensitivity to the cleavage between logical consistency and "real life" situations will become an important point of contention throughout the history of this subject. Most writers after Ramsey also feel uneasy, to a greater or lesser extent, as to what their attempts at logics of preference actually reflect. Some go to the extreme of insisting that consistency must become the sole criterion for such a logic, and the indeterminancy precipitated by chance is totally discounted. Others insist that such logics must reflect the "forms of life" which are peculiar to preference, and they avoid overall theories which attempt to distort the novelty of the subject to be characterized. Again, it is a tribute to Ramsey's perception that in this very first formal attempt, he foresees the sticking point which will limit not only his attempt but many which are to follow.

---

[8] Ramsey, F. P., *The Foundations of Mathematics and Other Logical Essays*, p. 186.
[9] ibid. p. 187.
[10] ibid. p. 177.

Ramsey's system for the measurement of degrees of belief begins with a minimum number of first assumptions.

First is posited the notion that the subject has certain beliefs about everything and that he acts in a way which he believes will bring about the best possible consequences.[11]

Second, it is supposed that the subject is capable of doubt, so that on certain occasions he would not be able to choose between options. Ramsey proposes to set forth axioms and definitions which deal with choices of the latter kinds. With the introduction of the notion of doubt, the groundwork is laid for developing formally the notion of "degrees of belief." For it will be seen that one's degree of belief in proposition p will be defined in terms of the option he must choose in securing a goal.[12]

Basic to his system is the notion of an "ethically neutral" proposition which is expressive of the case where, given two possible worlds differing solely in their respective truth value, both of these possible worlds are equal for some agent who must choose between them. This is to say that the agent would be willing to choose either one of the actions which would precipitate any of the two possible world states. Consequently, "ethically neutral" propositions are, as their name implies, propositions concerning states of affairs which are not valued in and for themselves, in any intrinsic sense. Such propositions refer to world states which the subject can decide to bring about without emotional compulsion. This is an important concept insofar as it fits well with Ramsey's analysis of partial belief, emphasis being placed upon the reasoned decision concerning alternative modes of action.

Ramsey's first definition is that of the degree of belief $1/2$ in an ethically neutral proposition. This is set forth as the case where the subject is indifferent between (1) and (2) as follows:

(1) $\underline{\alpha}$, $\underline{\beta}$
   is p is true   if p is false

(2) $\underline{\alpha}$, $\underline{\beta}$
   if p is false   if p is true

whenever the subject has a definite preference between α and β.[13]

---

[11] ibid. p. 176.
[12] ibid. p. 177.
[13] ibid. p. 179.

However, the subject *does have* some preference *between* α and β in each option.

The above definition is readily understood, once the notion of belief $1/2$ is explained. For Ramsey, full belief in a proposition is designated by 1, belief in its contradictory by 0, and equal belief in p and -p by belief $1/2$. Hence, belief $1/2$ in an ethically neutral proposition defines the agent's indifference toward options (1) and (2), while holding that the agent has a preference between propositions α and β within these options. Ramsey explains this definition as being roughly analogous to the situation where an individual is indifferent toward betting one way or another for the same stakes. Therefore, though the individual has a preference for winning the stakes, the options under which they are presented in no way enhance the stakes which are being considered; hence he is indifferent to the options themselves.[14]

Georg Henrik von Wright in his "Remarks on the Epistemology of Subjective Probability" renders this first definition by Ramsey perhaps too succinctly as follows. The degree of belief $1/2$ in an "arbitrary proposition p" (i.e. Ramsey's "ethically neutral proposition") is defined as an attitude of preference toward a simple option *combined with* an attitude of indifference in a conditioned option (i.e. an option whose total consequence is not known to the subject).[15]

The first definition is used to present a definition for the *difference* in value between two possible states *being equal to* the difference in value of two other possible world states having in common an ethically neutral proposition, say p. Ramsey presents this as:

1. p is an ethically neutral proposition, believed to degree $1/2$.
2. The subject has no preference for or is indifferent toward the following two options:
    (1) α if p is true, δ if p is false.
    (2) β if p is true, γ if p is false.
3. The indifference in value between α and β is equal to that between δ and γ, given the distribution of the values of p in (1) and (2).

   Hence, element 3. of the definition provides a means of expressing the sameness in value between possible world states.[16]

---

[14] ibid. p. 178.

[15] Von Wright, Georg Henrik, "Remarks on the Epistemology of Subjective Probability," in *Logic, Methodology and Philosophy of Science*, Proceedings of the 1960 International Congress (Stanford 1962) p. 332.

[16] Ramsey, F. P., *The Foundations of Mathematics and Other Logical Essays*, p. 178.

Again, von Wright's rendition of the second definition Ramsey offers is very much to the point. He interprets Ramsey as saying that the difference in value between two goods, G_1 and G_2, being equal to the difference in value between two other goods, G_3 and G_4, is defined as the holding of degree of belief $1/2$ in an (arbitrary) proposition p, *combined with* an attitude of indifference to a particular option conditioned by p.[17]

Ramsey is thus in the position to present the important idea that if possible world α is preferable to possible world β, then any other possible world having the *same value* as α is preferable to any other possible world, which has the same value as β. (For economy, 'α' will be used to designate both the world and its value.)

Ramsey's axioms for measuring value are straightforward, and the resulting theorems simply reflect the laws of commutation and of distribution. A representative sample of these are the following:[18]

(1) "There is an ethically neutral proposition p believed to degree $1/2$.
(2) If p [and] q are such propositions, and the option α if p, δ if not-p is equivalent to β if p, γ if not-p, then α if q, δ if not-q is equivalent to β if q, γ if not-q.*
(3) If option A is equivalent to option B and B to C, then A to C.
(4) If $αβ=γδ$, $γδ=ηζ$, then $αβ=ηζ$.

....

(8) Axiom of Archimedes"

*(In the above case the following definition is given: $αβ=γδ$.)

In the interest of brevity, it hardly seems necessary to give all of his axioms here. What is important is Ramsey's third definition, which von Wright identifies as the third crucial step of the entire presentation.

This is the definition governing the measurement of belief dealing with the subject's *belief in p given q*. This is seen as the case where one has indifference between the options:

---

[17] Von Wright, Georg Henrik, "Remarks on the Epistemology of Subjective Probability," p. 332.
[18] Ramsey, F. P., *The Foundations of Mathematics and Other Logical Essays*, pp. 178-179.

(1) α if q is true, β if q if false.
(2) γ if p is true, and q is true, δ if p is false, and q is true β if q is false.

The definition of the subject's degree of belief in p given q is the following ratio:[19]

$$(\alpha-\delta) : (\gamma-\delta).$$

By examining the distribution of the value of p in the above proportion one finds that q is always true, since it is given and p may be true or false, according to whether is appears either in γ or δ.

Von Wright's summation of this final step is that a person's degree of indifference in a certain conditioned option is employed to define one's degree of belief in a conditioned option. Though his interpretation of Ramsey's definitions is laconic, his overall aim is to test the fundamental feasibility of the logic Ramsey presents. With characteristic precision von Wright argues that Ramsey's definitions are *circular* because they assume the very subjective probability they are designed to define. At the outset he questions the notion of "indifference" which is so central to Ramsey's definitions. Von Wright indicates that there are at least two senses of the idea of indifference which must be considered. First, unreasoned indifference is the case where the individual agent is just too ignorant to choose between a set of options. Thus, his indifference is really the result of either lack of comprehension of the situation which is presented to him, or it is the result of stupidity. On the other hand, there is *reasoned indifference*, where the agent can choose or not choose between options, knowing that whatever course he takes will be to the same effect. The latter kind of indifference is based upon knowing the likelihood of certain results occurring. Hence, it is knowledge based upon awareness of subjective probability, i.e. an expectation as to the effect of an intended action.[20]

Von Wright observes that in the case of reasoned indifference, the possible effects stemming from choosing a future course of action are not seen in terms of how they each stand in a discrete *manner, but as they result accumulatively*. This accumulative aspect must play a part when considering the equality or inequality of value between two possible effects, a notion which cannot but figure prominently in any definition dealing with differences in value among two possible options, i.e. Ramsey's second definition. For it makes lit-

---

[19] ibid. p. 179-180.
[20] Von Wright, Georg Henrik, "Remarks on the Epistemology of Subjective Probability," p. 334.

tle sense to talk about equality or difference between effects, where the effects from options are conceived separately from each other, i.e. serially. Rather, in von Wright's view reasoned assessment requires that the agent expect a certain accumulation of effects to result, if he considers taking one option rather than another, within the context of the complexity of lived experience.[21]

If von Wright's insight is correct on this point, then it follows that the agent's deliberation upon any option requiring action must presuppose the expectation that certain effects repeat, and can be expected to occur as a total effect (i.e. accumulatively). However, if this much is granted, then the agent's very choice is itself being guided by a subjective probability, or belief in what may happen, which cannot be allowed to serve as part of the definiens of "subjective probability," e.g. contrary to what Ramsey's second definition says. This is the *point of circularity* which von Wright charges against Ramsey. Namely, one cannot define subjective probability in terms of values of future effects without also introducing subjective probabilities into the definition.[22]

One can only speculate as to why Ramsey did not consider the accumulative aspect of options in his analysis of preference. Perhaps his view of states of affairs and their effects was governed by his criticism of Keynes' view on belief in the truth of the conclusions of inductive arguments. It will be recalled at the outset of his attempt how Ramsey is attempting to argue that one cannot treat the conclusion of an induction as simply the result of the accumulation of the truths of its supporting premises, i.e. confirming instances. Ramsey is deeply concerned with the character of such a conclusion *as it is believed* by someone. In other words, he wants to describe the nature of the relation between believer and the proposition believed (i.e. the inductive conclusion itself). Ironically, it may be that in radically shifting away from Keynes' own "accumulative" position, he is forced to the opposite extreme of looking at propositions generally as atomic entities, having a fixed meaning, and becoming in a very structured way "objects" of belief. In fact, Ramsey alludes to the work of the early Wittgenstein to explain his conception of the ethically neutral proposition being fundamentally an atomic proposition along the lines considered in the *Tractatus*.[23] Hence, though the precise reasons for Ramsey missing the point on accumulative effects may never be known, it seems reasonable to suppose that his recoil from Keynes' view that inductive conclusions are believed simply because of the combined effect resulting from the truth of their supporting

---

[21] ibid. p. 334.
[22] ibid. p. 334-335.
[23] ibid. p. 165.

instances, together with his adoption of a view of language set forth in Wittgenstein's *Tractatus*, were the possible causes of his highly serialized conception of propositions expressing states of affairs. However, the flaw is sufficiently profound to cast doubt upon the viability of Ramsey's logic of preference. Moreover, Ramsey's repeated unease on the point of what his logic of preference reflected in the reality of everyday life may have been the result of his finding a distinct disintegration within the logic itself. This in turn may have been a result of the fact that he found himself forced to use a confining psychological theory to initiate his logic, which he admitted was generally discarded, though it was the only heuristic device which permitted the launching of his formalization of preference.

Von Wright's own solution to the problem of analyzing subjective probafjbilities is to deny that there really is a dichotomy between subjective and objective probability, as seems to be assumed throughout Ramsey's presentation. Rather, von Wright speaks of the *subjective significance* and the *objective significance* of assignments of probability. What one is dealing with here is the manifestation of certain beliefs and expectations in the case of the subjective significance of probability. On the other hand, the facts, if there are any, which justify the assignments themselves are the objective significance of probability. Though von Wright's views will merit their own critical assessment further on, his careful evaluation of the Ramsey line is clearly meritorious. Von Wright's realization that Ramsey is dealing with a debilitating subject/object split comes to the heart of the difficulty with the Ramseyan system.[24]

However, within the perspective of history, Ramsey's account stands as a brilliant beginning. His insight into how practical action forms the conceptual basis for measuring degrees of belief is not clearly anticipated anywhere else, save perhaps in the theoretical observations of Charles S. Peirce.

Ramsey's most penetrating insight is perhaps best illustrated by his observation that the procedure of extending the logic of consistency to the analysis of preference produces the effect of creating a distance between the logic of preference per se and the "real life" contingencies which influence actual preferences. This is to say that there seems to be operative an opposition or tension between what is normally taken to be the intension of preference and its extension, in that the more the consistency of preference is emphasized analytically, the more the extension of preference, i.e. its spatiotemporal aspect, is seen to recede from the surface features of the logic itself. This is apparently

---

[24] Von Wright, Georg Henrik, "Remarks on the Epistemology of Subjective Probability," p. 336.

the price which is demanded by the very first premise Ramsey begins with, namely, that "mathematical expectation" will form the basis for determining measurement in degrees of belief. Consequently, at the very outset, this mathematical requirement generates the condition where the ideal of "rational choice" does not necessarily reflect the richness of "lived experience". Though the mathematical picture is intended to capture something of the physical universe it measures, the fit is never precise, and the novelty of the real world seems always to remain elusive.

This very same concern with the interface of theory and reality within the context of the analysis of preference, which is first voiced in Ramsey's paper, will become a nagging problem throughout the history of the subject. The problem is recognized again after twenty five years by Kenneth J. Arrow in his classic book *Social Choice and Individual Value*. Though Arrow's work deserves a section of its own, his allusion to the difficulty in measuring utility for the purpose of setting forth preference scales provokes the following pertinent remark: "... If we look away from the mathematical aspect of the matter, it seems to make no sense to *add* (my italics) the utility of one individual, a psychic magnitude in his mind, with the utility of another individual ..."[25] Here Arrow is sensitive to the reconstructive role which mathematics plays in dealing with the human element which choice selection involves. It is interesting to note how he finds the "adding" of utility to be a process which rings hollow within the interpersonal dimension of experience. Thus he sees quite clearly, as Ramsey did before him, a cleavage between the mathematical model and the "form of life" which preference manifests in ordinary discourse. Earlier, he begins by noting that the requirement of consistency, wherein if a is preferred to b, and b to c, then a must be preferred to c, is something which cannot be insisted upon in everyday situations. The novelty of random events cautions against looking at theory-building as an end and not as a means. This is the position which Arrow adopts throughout his book, and his stated aim is to arrive at a means whereby one can discover patterns of uniformity "in a certain part of reality".[26] However, these uniformities are not to be drawn because of strictly "logical necessities" but because of the reality of the situation which exhibits them. Thus, for Arrow, the theory must be built around the facts, and not conversely. He therefor begins with the view that though one can have "adding" for an interval and/or normalized scale, the situation is simply not the same where the scale is ordinal.

---

[25] Arrow, Kenneth, *Social Choice and Individual Values*, John Wiley and Sons, New York, 1966, p. 11.

[26] ibid. p. 21. See footnote 18 on this page.

In essence, it can be said that whereas Ramsey put consistency as the primary requirement for a theory of preference, and then lamented how far away from reality he was forced to go by such a procedure, Arrow sees the same procedural problem which Ramsey saw but elects instead to work within the facts and to allow *their* regularity to manifest itself. Whereas Ramsey considers the difficulty of distancing which theory construction seems to impose on the subject at the *end* of his discussion, Arrow begins his work with the clear understanding that this danger exists, and thus rational preference ranking must be sensitive to what the facts demand.

II. *Davidson, McKinsey and Suppes on Logical Consistency and Preference Ranking*

While Arrow's work takes a different direction, an effort contemporary with his seeks further the same bent of inquiry initiated by Ramsey's *Truth and Probability*. This is the important paper by Davidson, McKinsey and Suppes, "Outlines of a Formal Theory of Value, I", published in 1955. In this study the authors face the issue of the relevance of a mathematical model to the reality of the physical universe. They advocate a proposed axiomatization which eliminates the need to "... depend on the inadequate and inflexible resources of ordinary language. ..." Furthermore, the axioms of preference they set forth are designed to avoid the frustration of explaining the fundamental concepts of value theory apart from a coherent system.[27]

Thus, like Ramsey before them, they place an extra premium on the logical consistency of their axiomatization of preference, and view their calculus removed from the "inflexible resources of ordinary language." The departure from Arrow is striking in that the abstract mathematization of preference is seen as an ideal rather than as an obstacle to achieving an accurate characterization.

Davidson, McKinsey and Suppes argue that those who oppose the possibility of the measurement of preference hold a too restrictive view of such measurement. Critics assume that since such measurement must involve the attachment of units to nonmaterial things, e.g. desires, wants, etc. it cannot allow for the "normal" determining of mathematical ratios among actual entities. Thus, they find Davidson's procedure overly simplistic, to the point

---

[27] Davidson, McKinsey and Suppes, "Outlines of a Formal Theory of Value, I," p. 142.

where what is taken to be the object of measurement gradually fails to attain signification within the logic itself. Davidson argues against this restrictive view of the matter. For example, the measurement of an apparently concrete thing like longitude is based upon the arbitrary selection of the zero meridian and the unit of longitude. In essence, the determination of locations on the earth is the result of a conceptualization of the globe which is mathematically abstract.[28] Davidson *et al.* urge that this is also the measuring of a physical thing, yet in a highly abstract and totally arbitrary context. Surely, the extensive criterion which critics of the measurement of preference demand is no more in evidence or even necessary in the determination of longitude. Here a more flexible sense of measurement is operative than that of the direct measurement of objects. In addition, the determination of temperature also cannot be measured in strictly extensional terms, yet it too is a physical phenomenon. Consequently, one must be cautious in denying the possibility of measurement for preference on the grounds that it does not lend itself to the kind of extensive measurement found in the surveying of physical things. To underscore the point, the authors cite the four types of measurement which are operative in the analysis of most physical phenomena: absolute scale, ratio scale, interval scale, and ordinal scale. Though these various mode of measurement are presented in the order of descending strength, Davidson claims that preference can be measured at least in terms of the interval scale. Hence, he will be advocating a means whereby preference can be measured in a way similar to the determination of time and longitude.[29]

It is vital to note that fundamentally these authors are really opposing those who want to discount the possibility of the measurement of preferences on the grounds that they are inaccessible psychological dispositions. Davidson, McKinsey and Suppes are thus defending two important points: (1) that there is a legitimate way to conceive of preference in totally systematic terms, involving primitives, rules of inference, and axioms, and (2) that the notion of measurement is broader than is ordinarily realized. The first point is well within the conceptual viewpoint first set down by Ramsey. The second point constitutes a departure from Ramsey's view, in that it amounts to a refinement

---

[28] ibid. p 150. (The critics to whom Davidson and his associates direct their comments are writers such as R.B. Perry and C.I. Lewis, who question the very tenability of attaching magnitude to nonextensive entities, such as pains, pleasures and desires.)

[29] ibid. p. 150.

of the conception of the instrument of mathematics as an analytical tool. There is no significant discussion in Ramsey's account as to the different types of measurement, as one finds discussed by these authors.

In passing, some assessment is needed of the defense Davidson and his colleagues are making for their specific approach. Particularly important is their view concerning the flexibility of mathematical measurement, since it appears to lead to an argument which is inconclusive for their cause. This is to say that to argue from the measurement of longitude to the measurement of preference seems highly elliptical. The measurement of longitude clearly deals with an abstract conceptualization of the earth. The arbitrary selection of the zero meridian in no way influences the result of the determination of longitude, once selection of the point of reference has been made. If these authors are contesting those who want to argue against the "precision" which such a logic (of preference) can have, then their argument from the measurement of longitude hardly touches the objection which seems to manifest the reservations of their critics. For in view of the novelty of human action and the difficulty in predicting its effect, how can one entertain a *logic* of preference? This seems to be the question recurring throughout the literature of this issue, which question is not resolved by alluding to the kind of measurement involved in the determination of longitude. Whether the zero point is chosen arbitrarily in no way influences the *results* which are dependent on that choice. This is to say that *arbitrariness* in the choice of the zero meridian does not influence the exactness of the longitude determinations after that choice has been made. As long as the rules for measuring such things are correctly *followed*, precision in measuring longitude is assured. However, the stability of the reference point which is a prerequisite for the measuring of longitude is absent in the case of the logic of a preference. In the latter case, one is dealing with activities having a uniquely human face, and as such they cannot be confined to logical consistency without the cleavage between it and its object manifesting itself. Thus, the argument Davidson and the others use for urging that common arbitrariness in selecting the starting point for measuring a phenomenon like longitude in no way supports their own thesis that preferences can be measured.

Davidson's response to the above is to state quite emphatically that those who oppose the possibility of the measurement of preference simply fail to understand that attaching a quantitative value to a preference option must be understood and appreciated in terms of a coherent theory of measurement. This theory is had by the axiomatization of conditions "... imposed on a structure of empirically realizable operations and relations. ..." It is complete if it can be shown that the structure "... satisfying the axioms is isomorphic to a numerical structure of a given kind. ..." Such theories are not designed to give empirical interpretations to sentences which express the mathematical probability of an agent choosing a certain preference option. Rather, they provide an empirical interpretation for "... *assertions* (my italics) about preferences, choices or decisions. ..." The interpretation by the theory of the observed phenomena results from considering the *whole* theory, wherein understanding of the isomorphism noted above serves to explain the nature of the reference between number and empirical structure.[30]

A scrutiny of what Davidson *et. al.* are claiming reveals that there still persists tension between theory and reality. Interestingly, they admit that their axiomatizations are *imposed* upon a *structure* of empirically recognizable operations. Also, they recognize the need to have an isomorphism between the axioms and numerical attributions within the system, so that the system can be complete. Moreover, they refer to the empirical interpretations of "assertions" concerning preferences, etc.; presumedly these are assertions within the system itself. Thus, it seems clear that these authors are looking at the physical realm as constructed by the theory they are projecting. It is still not the physical world to which they are addressing themselves, but a logical construction thereof.

Consequently, for Davidson, McKinsey and Suppes, one can indeed find a way of measuring preferences within the circumscribed field of theory. For their critics, however, one cannot measure preferences, if measurement here requires prediction with perfect accuracy as to the selection of a particular choice option by a subject within the physical domain. It could be said that both are right and both are wrong, depending on how each side wants to extend the reference of its claims. If Davidson admits to the limited scope of his claim, then, within the framework of a theory of preference, his measuring is a perfectly sound procedure. His critics are also correct, however, insofar as they insist that any theoretical characterization of the actual universe, as it operates by chance and at random, is intended for physical things and cannot serve analogically as a model for the interpretation and measurement of preference. Basic to their claim is the idea that preference is a private state, and any attempt to externalize it in terms of a physical action theory simply distorts what is being spoken about. Moreover, they would also urge that since the

---

[30] ibid. p. 151.

"real" world of physical phenomena cannot be precisely circumscribed within any theory, such theories cannot be extended to capture the subtle intricacies of subjective states.

Thus within their own respective domains, both Davidson and colleagues plus their critics are justified in their claims. Ironically, the argument leveled by each against the other really do not speak to the issues both sides wish to maintain. It is conceivable that Davidson could accept the premises of his critics' position as the result of a necessary drawback of any attempt at theory construction, while still consistently claiming his own position on the measurability of preference as being inherently sound. By contrast the converse may be the case for his critics. This is to say that critics may consistently deny the possibility of measuring preference in any *real* context and refer to the kind of thing Davidson and the others desire as simply an approximation of what is essentially inaccessible. For this reason, both sides are operating within their own restrictive patterns of discourse.

Apart from the authors' efforts to make known the many possible senses of preference-measurement, their conception of the formal aspects of preference is based upon a commitment to the idea that the roots of "rational choice" are to be found in "logical consistency". Hence, their argument for the feasibility of the measurement of preference is not based directly on mathematical considerations, as will be the case with Nicholas Rescher in 1968. Rather, their attempt is geared toward underscoring logical necessity, making this the criterion for evolving a viable logic of preference. Their adoption of the standard of logical consistency is clearly in keeping with Ramsey's approach, while their noting the variety of different types of measurement becomes only a supporting issue.

They emphasize the importance of the above criterion at the start when presenting their definition of "rational preference ranking, or RPR", which can be quoted briefly in its entirety as:[31]

"Definition 1. *The ordered triple ‹ K, P, E › is a*
RATIONAL PREFERENCE RANKING
if and only if:

P1. *The relation P is transitive;*

P2. *The relation E is transitive;*

P3. *If x and y are in K, then exactly only one of the following: xPy, yPx, xEy*"

---

[31] ibid. p. 143.

In the above, 'P' is understood as expressing the relation of preferring 'K': a set of alternatives to be ranked in preference, and 'E' represents the relation of equal preference status, which is transitive and symmetrical. ('E' can also be interpreted as designating the relation of indifference).

The authors stipulate that this first definition is not intended to say anything as to when or under what circumstances one alternative is preferred or equivalent to another in the set. The definition is only meant to impose limitations on the *patterns* of sets of alternatives. These patterns are judged to be rational if they reflect the requirement of logical consistency. Hence where a is preferred to b, and b to c, then that a must be preferred to c is the result of rational reasoning manifesting itself after the application of logical truth is taken *normatively*. In the case of the aforementioned argument, to claim that *therefore c* is preferred to a would justify saying that given the two former preference, this would *not* be a "rational" preference. Hence, the point Davidson and his colleagues seek to make is that the criterion of logical consistency is being used as a means of clarifying what it means to say that one has made a "rational choice". This particular function of the criterion in no way determines what we *should* believe, think, or support.[32]

To the objection that circumstances do not remain the same, and what seems to be x at one instance, t_1, may turn out to be different in another instance, t_2, Davidson argues that this could be remedied by attaching a subscript to x, so that where x_1 is preferred to y, and given yPz, it can be said that zPx_1, since at time t_2 the situation changed and the preference ranking which held at t_1 no longer holds. Davidson maintains that the original definition remains unscathed by the objection, since it speaks only to the *formal* properties of the patterns of preference and not to the particular circumstances of preferences themselves.[33] Presumably, Davidson feels that changes in the conditions of preferring would not alter the essential truth of the kinds of definitions he and the others are seeking, within a ratio scale.

Davidson's attempt to anticipate the possibility of a problem is not convincing. Clearly, if time factors are considered and changing circumstances recognized, then the issue of the identity of states of affairs comes into play. And this cannot be ignored by saying that such matters do not concern the formal patterns between rational preference rankings involving the ratio scale. Significantly, an important part of Definition 1. deals with the equality of preference rankings, designated by 'E'. Yet it is very difficult to speak

---

[32] ibid. p. 144.
[33] ibid. pp. 144-145.

meaningfully of equality in this context and not introduce some allusion to what it is for preferred states of affairs to be identical, or at least similar in respects which are not purely mathematical. This point is brought out quite clearly by Hans Reichenbach in his *Elements of Symbolic Logic*. Discussing the formal definition of the identity relationship, Reichenbach observes:[34]

"... two symbols denote the same thing if any two corresponding sentences, which contain these symbols in corresponding places, have equal truth-values.

The significance of this definition should not be underestimated. It reduces identity to equality, namely equality of certain properties. Thus in order to apply the definition we must be able to know whether the signs used in the corresponding sentences are equal to each other, in the sense of the similarity relation between *tokens* (my italics) ...; furthermore, we must know what equality of truth-value is. ... Although we may be compelled to apply the concept of identity before we understand the definition, we need not say it. ..."

It is interesting to note how Reichenbach attaches the determination of equality to the definition of identity, so that the latter is understood in terms of the uniform substitution of equivalents, resulting in sentences having the same truth value as the original sentence. Thus, the identity relation deeply involves the role of particulars insofar as they come to be understood as *tokens* for specific denotations. In this context, one notices that 'E', which in essence expresses the relation of "preferential indifference", can also be considered in terms of its expressing a sameness relationship between x and y, in that neither is being preferred. However, as Reichenbach's remarks suggest, sameness or equality, or even the reciprocal relation which preferential indifference entails, reflects the notion of identity or at least equality in some *particular* respect. Consequently, the reciprocity conveyed by 'E' can only be a token relationship, not a type relationship. Thus, again, to insist that Definition 1., together with all the other definitions to be offered is entirely formal and separate from any considerations dealing with particulars, is to obfuscate the notions these definitions are designed to clarify.

If Davidson and his colleagues recognize the role of time and the alterations it can induce in circumstances generally, then the relation designated by

---

[34] Reichenbach, Hans, *Elements of Symbolic Logic*, The Free Press, New York, 1966, pp. 241-242.

'E' cannot be meaningfully universal or general in any sense and still remain of any practical use in calculating preferences. Granting this, it cannot be maintained that the identification of formal patterns of ratio-rankings among preferences say nothing about particular circumstances. Again, it is not difficult to see here the gradual reemergence of the same problem found in Ramsey's account, which resulted in the distancing of theory from reality.

The preference-ranking relation Davidson defines at the outset of his paper is admittedly inadequate for expressing the complexity of certain situations, where preferences exhibit richer possibilities than Definition 1. is designed to cover. He illustrates this by citing the case of the congressman who must form a decision as to the outcome of an issue in which he has an interest. The issue is that of a bill requiring federal subsidy to schools not practicing racial discrimination. Congressman Wright knows that as the bill stands it will not pass with the anti-discrimination clause attached to it. On the other hand, he knows that without the clause the bill has better than an even chance of passing. His third alternative is to withdraw the bill. The congressman sees his options as follows: (a) there is one chance in three for the full bill to pass as it stands, (b) the bill is assured passage without the anti-discrimination amendment, and (c) the bill is defeated or withdrawn.

Should he try for passage of the full bill and risk certain defeat? Davidson labels this Action 1. On the other hand, should he push for a weak bill and be assured of its passage? This Davidson calls Action 2..[35]

The above case illustrates that something more needs to be considered than just a strict preference-ranking as Definition 1. provides. Some means should be available so as to determine *how much more* the congressman values option (a) over option (b), and (b) over (c), or conversely. Davidson's proposal for solving this problem is to introduce the interval scale into his analysis and thus expand the definition of rational preference-ranking.[36] This should take care of accounting for the element of comparative success between diverse recourses for action. Consequently, in the case of the passage of the school subsidy bill, the congressman would reason in the following way. Option *a* above would be assigned a value of 2, *b* the value of 1, and *c* the value of 0. The authors note that these numbers express only the relative value of each option, and nothing more is intended in their use. They are just an indication of rela-

---

[35] Davidson, McKinsey and Suppes, "Outlines of a Formal Theory of Value, I," p. 142.
[36] ibid. p. 152.

tive "magnitude of difference". Thus Action 1. entails one chance in three in securing a, valued at 2, and two chances in three for getting c, valued at 0. Mathematically, Action 1. can be represented as:

$$\frac{1}{3} \cdot 2 + \frac{2}{3} \cdot 0 = \frac{2}{3}$$

It can be seen that as long as the ratio of the difference between a and b, and b and c, is less than 2, the congressman can be expected to take Action 2. If this ratio is greater than 2, then he will take Action 1. Most importantly, Davidson, McKinsey and Suppes point out that the weight which is assigned to each of the options by the congressman reflects his own opinion as to the probability of their occurrence.[37]

The ratio scale simply cannot handle the subtleties of the comparative element as easily as the interval scale. In recognition of this, Davidson employs the case of Congressman Wright's decision-making procedure to introduce the latter scale as it is used in the von Neumann- Morgenstern theory of preference, thereby endeavoring to strengthen Definition 1. However, without needing to devote attention immediately at this juncture to efforts by economists, it is both interesting and important to understand why, on purely philosophical grounds, Davidson, McKinsey, and Suppes are forced to the conclusion that the first definition of rational preference-ranking is inadequate.

Basically, in the example of the congressman, one sees that the options the subject is faced with do not allow for the relation of equality relative to preference status, as expressed by 'E'. In other words, the congressman cannot be said to prefer equally the passage of the bill with the crippling amendment, as well as the passage of the same bill without the amendment, and finally the defeat of the bill. Somehow the situation, as denoted by either a, b, or c, does not allow for the possibility of indifference among the options, which is what equality of preference status actually involves. Though the authors do not dwell on the precise reasons for the inadequacy of Definition 1., the flaw stems mainly from their unwillingness to accept the fact already noted, namely that if 'E' is going to have any functional use, it must reflect a token relationship and not a type relationship. Davidson wants his formalization of preference-rankings to be strictly formal. However, in insisting upon this, he and his colleagues find that their definition fails to have any practical use, in that particular occasions, such as that of Congressman Wright's decision, demand a specificity which the formal structure cannot express. They allude to the von

---

[37] ibid. p. 149.

Neumann and Morgenstern axiomatization as a way of introducing greater flexibility into their formalization. Yet, as will be seen, this maneuver is really a way of talking around the issue of extensionality, so as to remain on an entirely formal plane. In the end, the familiar caution is voiced by them concerning the limitation which their formalization of preference must admit when faced with the complexities of the physical world. The alternative axiomatization Davidson, McKinsey and Suppes propose is presented to suggest what the measuring of preference "might" be like within an interval scale. (It is interesting to note here the provisional nature of the remarks which the authors proceed to offer.) The strategy is to introduce a single new primitive into Definition 1., so as to accommodate the element of probability of occurrence surrounding the choice of alternatives the subject is presented with. This primitive function h has three arguments such that where x and y are in K, and where α expresses the probability not equal to 0 or to 1, (i.e., $\alpha = (0 < \alpha < 1)$), then $h(x,y,\alpha)$ is the combination where x has the probability and y has the probability $1-\alpha$.

The new definition, assuming the idea of an open interval (0, 1), emerges as follows:[38]

> Definition 2. The ordered quadruple ‹K, P, E, h › is a
> RATIONAL PREFERENCE PATTERN IN SENSE ONE
> if and only if for every x, y, and z in K and every α and β in (0, 1):
>
> H1. The ordered triple ‹ K, P, E › is an RPR (in the sense of Definition 1.);
> H2. $h(x, y, \alpha)$ is in K;
> H3. If xEy, then $h(x, z, \alpha) E h(y, z, \alpha)$;
> H4. If xPy, then $xPh(x,y,\alpha)$ and $h(x,y,\alpha) Py$;
> H5. If xPy and yPz, then there is a number γ in (0, 1) such that $yPh(x,z,\gamma)$;
> H6. If xPy and yPz, then there is a number γ in (0, 1) such that $h(x,z,\gamma)Py$;
> H7. $h(x,y,\alpha) = h(y,x, 1-\alpha)$
> H8. $h(h(x,y,\alpha),y,\beta) E h(x,y,\alpha\beta)$.

The intuitive validity of most of the above axioms is self-evident, and thus they hardly deserve comment. The ones requiring some exegesis are H4., H5., and H6. Axiom H4. simply states that where x is preferred to y, then x is preferred to any probability combination of x and y, and any probability-combination of

---
[38] ibid. pp. 152-153.

x and y is preferred to y. Axiom H5. and H6. state that where y is preferred somewhere between x and z, then there is a probability-combination of x and z which is preferred to y: H6., and another such probability-combination over which y is preferred: H5.

The authors go to great pains to illustrate the formal adequacy of Definition 2., and offer arguments showing how this provides at least the necessary conditions for the rationality of preferences "among alternative involving uncertainty." Their presentation is technically correct. For what they claim is simply the result of applying Definition 2. to certain hypothetical problems, thereby showing how it helps in arriving at a means for determining what counts as a rational preference-ranking within a circumscribed field.

Of far greater philosophical interest is their expression of misgiving on the consequences which seem to follow from Definition 2. Their reservations concerning the "explicative validity" of this definition leads them to offer even a third definition which attempts to cover the glaring shortcomings they fall upon. However, the objections they note emerge not from philosophical analyses they or others have performed, but from critical analyses which economists have made of the above axiomatization, particularly as it is evident in the work by von Neumann and Morgenstern. Oddly, these objections come to reflect a profound philosophical significance, though they are voiced in quarters other than those in which philosophical analysis prevails. This single fact perhaps more than any other illustrates the virtual absence of any philosophical discussion of preference *within the analytical tradition* from the time of Ramsey's writings in 1931 to the present work by Davidson and his associates. By contrast, economists deal with the problem of the mathematization of preference from the early 1920's onwards. In fact, a case can be made for saying that the rudiments for the formalization of preference in terms of attaching numerical units to certain states of preference (considered entirely within a subjective context) were suggested as early as Alfred Marshall's *Principles of Economics*. Thus, one has here the unexpected case of an emergent philosophical concern appearing in an area outside the discipline of philosophy proper, gradually working into the mainstream of philosophical activity. It is important to note as well that the only "philosopher" within the analytic tradition who actively pursues the problem of the formalization of preference, apart from Davidson and the others, is Herbert G. Bohnert, in his seminal paper, "The Logical Structure of the Utility Concept," published in 1954.[39] Unfortunately, Bohnert's ef-

---

[39] Bohnert, Herbert G., "The Logical Structure of the Utility Concept," in *Decision Processes*, edited by R.M. Thall, C.H. Coombs, and R.L. Davis, John Wiley & Sons, Inc., New York, 1954.

fort has not received the attention it deserves, and his critique of the von Neumann-Morgenstern axiomatization will prove most difficult to sidestep.

At the present juncture it suffices to consider Davidson's own inspection of this second definition and the problems he finds economists have voiced as to its explicative adequacy. First, Davidson notes that as the definition now stands it says nothing as to how the element of risk may influence the value of an alternative. This is especially true of axioms H3, H4 and H8 above. For example, there may be the case where the subject would on the one hand prefer tossing a coin to determine whether he or his friend will pay for a meal, *even though* he wants to pay for the meal, rather than to have his friend do so. In essence, one can see how Definition 2. says nothing about this hierarchic structures of preference, which is an added dimension of the entire matter dealing with preference-rankings. Moreover, the example illustrates how the alternatives the subject is faced with are made problematic through interjection of the possibility of risk.[40] This means that the tossing of the coin serves to alter the original alternative fundamentally preferred by the subject, e.g. paying for the meal of his friend. For it may turn out that if his friend wins the toss, the friend will pay. Consequently, one sees how introducing the risk factor produces results which may conflict with the consistency of the preference pattern, within a given context. This serves to work against what Davidson and his associates find to be the foundation of their attempt to present a theory of value, which was seen to be the requirement of logical consistency.

Secondly, Davidson observes that the definition imposes on K, the ordered triple, the requirement that it must contain finite probability-combinations of all initial alternatives, and therefore alternatives of an infinite number. Here Davidson and his colleagues make perhaps the most philosophically penetrating remark in the entire paper. Quite candidly they admit: "... There is a reasonable doubt, at least, what empirical interpretation can be given to the assumption of an infinite number of alternatives".[41] This comment reflects once more an undercurrent of concern with the way their theory relates to the actual world. For it is an objection which speaks to the soundness of the proposed conceptualization. Raising the question in the context of trying to give an "empirical interpretation" again betrays to some extent a mistrust of a totally formalistic approach, the ramifications of which may not always be either entirely consistent or relevant in specific cases.

---

[40] Davidson, McKinsey and Suppes, "Outlines of a Formal Theory Value, I," pp. 155-156.
[41] ibid. p. 156.

The remedy proposed by these authors for the second deficiency is to introduce a new variable, so as to keep separate assumptions concerning the number of initial alternatives, from assumptions dealing with the number of finite probability combinations. Thus 'T' is presented as a new primitive expressing a quaternary relation. Actually, Definition 1. remains unchanged except for T, whose interpretation runs as follows: If x, y, and z are in ordered set K, and α is a probability in the closed interval (0,1) (i.e. $0 \leq \alpha \leq <$), then T (x,y,z,α) if and only if y is not preferred to x; z is not preferred to y; and "the alternative consisting of x with probability α and z with probability 1-α is *equivalent in preference* (my italics) to alternative y.

By way of explicating the upcoming definition, it is important to clarify first the function of 'T'. 'T' serves to specify the range of possible probability-combinations within a closed interval. Consequently, it maintains an ordered ranking between x, y, and z, so that there is no case where y can be preferred to x, or z to y, and it insures an equality between the probability of x in conjunction with the probability of z, in relation of y's. Thus T serves the planned purpose of preventing the possibility of an infinite number of alternatives. Hence the new definition appears as:[42]

Definition 3. ‹ K, P. E, T › *is a*
RATIONAL PREFERENCE PATTERN IN SENSE TWO
if and only if for every x, y, z, and w in K and every α and β in (0, 1):

T1. ‹K, P, E› is an RPR (in the sense of Definition 1.);
T2. If xEy and yEz. then $T(x, y, z, \alpha)$;
T3. If xPy or zPx, then not $T(x, y, z, \alpha)$;
T4. If $T(x, y, z, \alpha)$ and xEw, then $T(w, y, z, \alpha)$;
T5. If $T(x, y, z, \alpha)$ and yEw, then $T(x, w, z, \alpha)$;
T6. If $T(x, y, z, \alpha)$ and zEw, then $T(x, y, w, \alpha)$;
T7. If xPz, then yEz if and only if $T(x, y, z, 0)$;
T8. If xPz, then xEy if and only if $T(x, y, z, 1)$;
T9. If xPz, not yPx and not zPy. then there is a unique γ in (0,1) such that $T(x, y, z, \gamma)$;
T10. If xPz, yPz and zPw, and any two of the following, then the other two:
$T(x, y, w, \alpha)$
$T(x, y, w, \underline{\alpha})$
 β
$T(x, y, w, \beta)$
$T(x, y, z, \underline{\alpha-\beta})$
  $1-\beta$

---
[42] ibid. p. 156.

This last definition, with its consequent axioms, simply expresses the probability-distribution for x, y, and z, within ordered set K, given the four place relational predicate T. As such it is philosophically uninteresting, and represents an attempt to interface values and probabilities within a highly restricted context, such as that found in economics.

Moreover, as Davidson notes at the conclusion of the essay, the purely mathematical interpretation of probability cannot be employed to define or totally explain psychological expectation. Hence, apart from what can be said concerning the probability of one event occurring rather than another, this by itself cannot completely explain a subject's personal belief or certain expectation that an occurrence will come about, despite the "odds".[43] With this, Davidson, McKinsey and Suppes are granting that subjective probability defies characterization in terms of the logic of probability, and therefore one is forever stopped from completely capturing the logic of preference in this way.

Though he recognizes the importance of the approach suggested by Ramsey, where subjective probability is treated within a behavioral context of action, Davidson still seeks to replace it with a conception of rational consistency which is more precise, in that it adheres to strictly mathematical requirements as proffered by von Neumann and Morgenstern. Consequently, it can be said that Ramsey's attempt is not extended by what the Davidson paper offers, in that Ramsey never chooses to lose sight of the need to account for the psychological element in preference. On the other hand, Davidson and his colleagues forego any consideration dealing with the manner of expressing the subject's private mental life. For these reasons, Ramsey's attempt can be said to have a different thrust than that of Davidson's. Basically, what the latter offers is a mathematico-economic analysis within a context of consumer demand and price determination, rather than an analysis aimed toward describing subjective probability in terms of the ratio of alternative modes of action.

III. *Herbert G. Bohnert on the "Eventhood" of Frequency Probabilities*

It is at this point, in light of the complete presentation of Davidson's adaptation of the von Neumann-Morgenstern definition of rational preference ranking, that Herbert G. Bohnert's criticism is most appropriate. Bohnert's work is outstanding in its penetrating facility, as it examines this theory's consistency, in the light of a theory of events.

---

[43] ibid. p. 159.

Bohnert begins by focusing upon explaining the form of statements dealfjing with concepts which are presumed to be clear. He reasons that if a concept is fully comprehensible, then the form or structure of the sentence expressing this concept should be specifiable in some simple way. Thus, in the context of theoretical discourse, one should have no difficulty explicating the structure of sentences expressing defined concepts within that context. However, the above point concerning the specification of sentential form becomes bothersome where the results of empirical measurement come into play, for then the theoretical context and its empirical counterpart must be reconciled in some way.[44] Bohnert recognizes this as the essential problem which underlies attempts to adopt theories concerning utility measurement, such as the von Neumann-Morgenstern theory.

Bohnert investigates a basic tenet of this theory which he renders as: "The utility of E at time t is u utiles (or units)."[45] He sees this in terms of the central question dealing with the nature of the entities over which 'E' ranges. Can it be consistently maintained that the probability-combinations of events are themselves events? One sense of event is that which is designated by a sentence. This is what Carnap terms the "proposition" of the complete sentence. Propositions, as states of affairs, allow for the ascription of truth values. For example, the sentence expressing that a certain coin is tossed a certain number of times can be verified so as to determine whether or not *what* the sentence says is true. On the other hand, Bohnert recognizes a different sense of "event," where event is considered in terms of its being a propositional function. This involves saying that the event the expressions refers to is *not a specific state*, in a determinate place and time, which would make possible the attribution of truth value for what is being considered. For example, *a coin toss* is not expressive of a proposition in the sense of a state of affairs designated by a complete sentence, but it is a sentential function which *predicates*, e.g. "x is a coin toss". It makes no sense to attach truth values to propositional functions, though they can be said to express greater or lesser probability in some cases, once instantiation occurs. The usefulness of propositional functions is that they provide a means of expressing the *likelihood* of an occurrence, which in the von Neumann-Morgenstern theory is treated as a frequency-probability. Where proper instantiation takes place, then the schemata expressing the propositional function are seen to yield propositions, involving probabilities. However, this is not to say that all propositional functions express only frequency-probabilities.

In the von Neumann-Morgenstern approach, events are considered in terms of their being propositional functions. This appears to be in keeping

---
[44] Bohnert, Herbert G., "The Logical Structure of the Utility Concept," p. 221.
[45] ibid. p. 224.

with the way economists generally work with probabilities so as to express relative frequencies. On a closer look, however, Bohnert finds that some interpretations of this particular theory assume that the events referred to in the basic class, from which the probabilities are projected, are also "... conceived as relative to a standard moment ... in the immediate future. ..." Thus according to these interpretations, at the lowest level, preference is seen as holding between propositions, and not between propositional functions. For Bohnert, the two senses of "event", e.g. that which is referred to by a complete sentence and that which involves a propositional function, are different and require that a sharp dichotomy be maintained between the two. Interpretations which confound the two senses of event, such as is the case in the above account of the von Neumann-Morgenstern utility theory, cannot but misrepresent the theory at hand. Basically, it is *only* events as propositional functions that this theory deals with. There simply is no reference to "facts" in theories which deal with relative frequencies, and thus the former sense of event cannot figure into the interpretation of such theories.[46]

The difficulty in trying to visualize the concept of event which is operative in this theory stems from the fact that the very notion of preference, as some sort of instantiation within a propositional function, is taken as a primitive. At no place does this theory introduce an explicit definition of what preferences are. Rather, it characterizes preferences implicitly by formulating a theory on the basis of preferences. The *modus operandi* which is favored proceeds along the lines of showing what preferences are by seeing whether the theory itself is adequate for explaining some intended purpose. Instructive insights such as these appear in Richard M. Martin's *Belief, Existence and Meaning* and they clearly outline how in the von Neumann-Morgenstern utility theory, the presupposition is that events are defined recursively, and not explicitly or extensionally.[47] Thus, the notion of preference in this theory is couched in terms of relations of relative frequency, which trade on mathematical probabilities.

Differentiating between the two sense of "event" has further telling consequences for understanding the theory under discussion. For example, given the hypothetical case of a subject who has a choice between two elements, A and B, with the probability p and -p respectively, if A and B are propositional functions, given that the probability is a frequency-probability, then the proba-

---

[46] ibid. p. 225.
[47] Martin, Richard M., *Belief, Existence and Meaning*, New York University Press, New York, 1969, p. 99.

bility of A must be a frequency probability relative to some other probability, designated by C. This is to say that any frequency-probability can be relativized to some other frequency-probability, which serves as a context for the former. In other words, to say that there is a frequency-probability of A, means that there must be the possibility of a probability frequency for the subject to decide whether to act relative to the option probability A provides. Bohnert uses the example of a subject faced with the situation of betting at a game of roulette, to illustrate the relation between A and C above. If betting that the ball will fall in a particular slot on the roulette wheel at such and such a time constitutes the factor of probability function A, then C is the probability of the subject accepting a promise (the roulette keeper's word that he will spin the wheel when the bet is placed), which will be realized at some future time. Furthermore, C must be an occurrence which is always followed by an occurrence of A or B. Moreover, Bohnert stipulates that the relative frequency of the joint occurrence of A and C, in relation to the occurrence of C, must be p. Leaving aside questions dealing with the infinite regress possibility of frequency-probabilities within successive contexts, Bohnert observes that the von Neumann-Morgenstern approach must be interpreted as being concerned with events only in the sense of propositional functions suitable for frequency-probabilities, and not events as the denotation of Carnapian propositions. Thus, not only must one construe A and B as probability-frequencies, but for consistency, C must be considered in terms of a probability frequency as well.[48]

Bohnert's analysis demonstrates that the theory under consideration works in the highly restricted context of frequency-probability. Its advantage over simpler ratio scale analyses, as was pointed out independently by Davidson, is that it enables one to express magnitudes of choice much more easily, thus freeing one from seeing preference solely in terms of numerical ordering. However, the drawback for Bohnert is that, in very general terms, this utility theory cannot be made to express the choice situations in the "real world." This is to say that in actual cases one does not have the "pre-set random device" implicit in choices between probability combinations in actual cases.[49] Usually, one's actions are influenced by habit, prejudice or limited information. Bohnert's rejection of the theory of relative frequency as a means of analyzing the nature of preference is perhaps more fully realized in an earlier work by William Kneale, entitled *Probability and Induction*. Discussing the many rea-

---

[48] Bohnert, Herbert G., "The Logical Structure of the Utility Concept," p. 226.
[49] ibid. p. 225.

sons for rejecting the frequency theory, Kneale notes the following among his many observations:[50]

"A second reason for rejecting the frequency theory is that it does not enable us to understand why it is rational to act on considerations of probability. Let us suppose for the sake of argument that the assertions of the frequency theorists are true. What is their relevance to the situation of a man who knows of something only that it is α but has to decide whether or not to act as though it were β? The fact, if it is a fact, that the limiting frequency of β-ness in the collection of α things is greater than $1/2$ seems to have no *direct bearing* (my italics) on the particular problem, for it is concerned neither with the individual thing as such nor yet with the character of α-ness which it is known to have, but solely with the way in which β things happen to be distributed in an infinite succession of α things. It is not necessary to labour this point, for many frequency theorists have themselves stated it clearly. Thus von Mises writes: 'We have nothing to say about the chances of life and death of an individual, even if we know his condition of life and health in detail. The phrase "probability of death", when it refers to a single person, has no meaning at all for us.'[1] Since practical decisions, e.g. of the managers of insurance companies, always refer to individual cases, this statement, taken alone, would imply that considerations of probability can never be of any use in practical affairs. But frequency theorists usually go on to say that their theory is nevertheless a trustworthy guide in certain field of practice, such as insurance, where we have to do with many instances of a kind;[2] it is necessary to examine their argument in some detail."

"... But it has been maintained that, if a man acted consistently on probability rules as defined by frequency theory, he would inevitably have more successes than failures in the long run. According to this argument the rationality of acting on considerations of probability in a particular case is derivative from the rationality of the *policy* of acting always on these considerations in cases of the same kind. But what is meant here by 'the long run'? The developed frequency theory does not allow us to predict with certainty the frequency to be found in any finite succession of trials, however long. ..."

---

[50] Kneale, William, *Probability and Induction*, Oxford University Press, 1963, pp. 164-166.

As in the case of Bohnert's criticism, Kneale also reveals the existence of a persistent cleavage between theory and the world. This is precipitated because analysts take a theory from the circumscribed domain of economics and project it into the broader area of philosophy, whereupon a tension develops between the requirements of the theory and the "reality" which the philosophers conceive. The omniscience which the von Neumann-Morgenstern theory assumes becomes a point of serious contention in the philosophical account of the rationality of preference. The possibility of choice between probability combinations, which is taken for granted by this economic theory, becomes an open issue when dealing with the alternatives the philosopher is faced with in trying to explain human action.

Chronologically, Bohnert's work appears prior to that of Davidson's. Yet its relevance to the implications of the latter's adoption of the von Neumann-Morgenstern theory is instructive. Bohnert himself does not bring out most of the ramifications of his own position. Especially important is the fact that where he discerns between the two senses of "event", he is making a significant *linguistic* distinction. This is to say that in speaking of "event", expressed as that proposition which is designated by a sentence, and "event" as the function of a proposition, he is distinguishing between two distinct ways by which signification accrues from a sentential form. Basically, it represents the differentiation of meaning resulting from sentential reference, and meaning resulting from instantiation into a sentential function. The utility theory under consideration deals with preferences in terms of their being frequency-probabilities, and thus the sense of event it presupposes is that which results from instantiation into propositional functions. On the other hand, "event" expressed as that which is designated by a sentence, suggests that there is an isometry between the sentential expression itself and the state of particular things it refers to. In this latter case, an expression of preference achieves its particular signification because it iconifies particular facts. Without distorting the original Carnapian insight, it is only in this context that one can speak of a "true" preference in the sense of having the possibility of verifying the extension of subject, object of need, observable action, time span, etc.

With the vital clarification of the above two distinct senses of "event" Bohnert brings forth, one can see more precisely the reasons why Davidson and the others register disaffection with their adaptation of the aforementioned theory. The problem arises because they are dealing with a conception of preference which is predicated upon the notion of frequency-probability. However, as philosophers who are interested in the clarification of ideas, they claim to be concerned with a general view of preference, upon which logical analysis can shed light. This conception of preference most likely has its origins in ordinary discourse, since it is inconceivable to think of Davidson and his associates as attempting to investigate an essentially metaphysical view of preference through the use of a utility theory found in economics.

One pursues the analysis of ideas in ordinary discourse either in terms of their extension (i.e. reference) or in terms of their intension (i.e. context). The former involves aspects of the analysis suggested by Carnap. The latter requires an open-ended view of language analysis, as found in the works of G.E. Moore and the later Wittgenstein. Davidson and his associates forego the latter kind of investigation because they see it as yielding imprecise results, that is, results which are not open to a systematic treatment allowing for the introduction of mathematical factors. On the other hand, they also do not subscribe to a strictly denotative mode of analysis because they regard this as "inflexible." As a result of disowning both of these two options, Davidson, McKinsey and Suppes are left with a conception of preference involving frequency- probability, as developed in the von Neumann-Morgenstern approach. Yet they cannot argue that this alone is the only valid conception of preference, since they are sensitive to the fact of how far short this particular notion falls from the general meaning of preference, as found in ordinary contexts.

Moreover, the very fact of their comparing preference as a frequency-probability with the idea of preference as manifested in some ordinary language context suggests that they consent to an underlying conception of preference which however dimly intuited, is broader and more complex than that found in the above utility theory. Interestingly, though this broader sense of preference may be vaguely seen by these authors, its articulation sententially must be *at least* susceptible to an extensional mode of analysis which, as has been shown, commits one to a sense of event in the denotative mode. The point here is simply that any sentential expression of ordinary discourse can be shown to have some basic element of extensionality or reference. Even in the very unusual cases of sentences having to do with nonreferring expressions, one can indicate their extensional structure easily, as Bertrand Russell pointed out in the essay "Description." Consequently, since event as propositions are evidently presupposed within this broader sense of preference, there is tension with any theory in which the underlying conception of event is defined recursively as a frequency-probability. In essence, one has here the same point Bohnert brings out above, regarding the need to be consistent with the way one interprets the kind of event the von Neumann- Morgenstern utility theory works with. For these two senses of event are mutually exclusive. This is to say that there does not seem to be a way by which one can arrive at a notion of event as a frequency-probability from a conception of event as a specific state of affairs, designated by a sentential expression.

Here it is beneficial to recall the criticism von Wright leveled against Ramsey's conception of subjective probability. Von Wright spoke of the subject's expectation of the good which certain events will realize as an *accumulative* good, and not as a serial ordering of discrete goods. One could use this important insight and argue that frequency- probabilities have this same accumulative aspect, in that they involve the notion of the class of a number of events, and not the specific occurrence of a particular state of affairs. Hence, the former notion involves the accumulative effect resulting from repetitive occurrence. Though it can be argued that it is specific occurrences which go into composing the class of events known as a frequency-probability, still the value of the latter as an analytical tool within a utility theory is in its expressing a propositional function and not a specific denotation. In this way one can perhaps see more precisely the nature of the distinction between the notion of event as a proposition designated by a sentential expression, and an event as the function of a proposition. The former can be given a determination of truth or falsehood; the latter cannot be so characterized.

The initial stance by Davidson and his associates, where they close the door to the possibility of arriving at a conception of preference through the analysis of ordinary discourse, not only involves them in the difficult problem of trying to reconcile an apparently specialized notion of preference with a much more generally accepted conceptualization of this term, but it places into question any point of tangence between economic and philosophic analyses.

Bohnert's plan, on the other hand, is to sketch, in very general terms, a formal description of rational preference-relations, where logical probability replaces the notion of relative frequency. His strategy is to define within Carnap's theory of logical probability (i.e. *probability$_1$*) a notion of utility, which will provide a basis for a definition of preference. He sees logical probability as a generalization of logical implication. Thus if e logically implies h, then in any case where e is true then h must also be true. This is a way of saying that given the range of possible worlds e (i.e. state-descriptions within a system) is true, then h is a subset of e, and hence h is also true.[51]

---

[51] Bohnert, Herbert G., "The Logical Structure of the Utility Concept," p. 226.

By using logical probability, Bohnert hopes to avoid some of the difficulties encountered in the von Neumann-Morgenstern theory with regard to the characterization of preference. The notion of logical probability is not equated with that of logical implication, since the former is taken to be "weaker" than the latter. Bohnert is found to refer to logical probability as the "confirmation" or "support" that one conclusion follows from an antecedent. The point is clearly a reflection of Carnap's position in *Logical Foundation of Probability*. Essentially, Bohnert is following Carnap in saying that, in one sense, probability deals with the question of "... the degree of confirmation of a hypothesis *h* with respect to an evidence statement *e*, e.g. an observational report. ..."[52] In this respect it is fair to say that Bohnert sees the issue of the proper characterization of confirmation *h* by *e* as a logico-semantical problem, where all that is needed is to establish the nature of the connection between hypothesis and supporting statement(s) as the logical analysis of the meanings of both.

Bohnert presents his own solution to the problem of how to properly formulate the concept of utility, from which he seeks to arrive at a more suitable definition of the notion of preference. He begins by having 'e(x,t)' express the proposition of the subject x's *total* experience up to time t. Thus he goes on to define "proximate utility" as follows:[53]

"... *The proximate rational utility of proposition p for x at t with respect to time-span s equals u utiles* if and only if p is logically compatible with the total experience [e] of x at t and the estimate of his perspective happiness during s beginning at t on the basis of p together with his total experience at t equals u."

In symbolic terms, Bohnert renders the above definition as:

$U(p,x,t,s) = u = \Diamond (p.e(x,t) \cdot (E(H,x,t,s,p.e(x,t))=u)^*$

*'E' expresses the estimate of the hedonic element.
*'H' expresses the hedonic experience over time-span s.

With the above Bohnert goes ahead to sketch a formalization of rational preference, P, as:[54]

$P(x,p,q,t) =_{df} \Diamond(p.e(x.t)) \cdot \Diamond(q.e(x.t)) \cdot {}^*U(p, x, t) > U(q, x, t))$

---

[52] Carnap, Rudolf, *Logical Foundations of Probability*, The University of Chicago Press, 1950, pp. 19-20.
[53] Bohnert, Herbert G., "The Logical Structure of the Utility Concept," p. 226.
[54] ibid. p. 228.

Thus subject x refers proposition p to proposition q at time t if and only if the conjunction of the logically possible propositions p and q holds with the utility value of p being greater than that of q.

In assessing Professor Bohnert's contribution to the development of the logic of preference, three distinctions must be maintained as to what he endeavors to discuss. First, his paper is geared mainly to the examination of the concept of "utility", as reflecting the von Neumann-Morgenstern theory. Second, his suggestion — that the concept of logical probability, in the sense of Carnap's *probability*$_1$, replaces the notion of relative frequency in defining the concept of utility — initiates a change in the way this notion has been handled by investigators interested in statistical mechanics. Third, and a point more directly related to the present discussion, are his insights into the formalization of the notion of preference as derived from his definition of "proximate utility." Though discussion on the last point cannot be totally separated from what Bohnert says in his definition of proximate utility, emphasis will be placed on the ramifications of Bohnert's effort at evolving a definition of preference from his notion of utility.

It must be granted that of the writers concentrating in the area at this time, only Bohnert analyzes preference in a logico-semantical context. In doing so, he is trying to approach a definitional schema of preference which reflects how this term functions in ordinary discourse, though clarified extensionally and made precise through logical analysis. His use of logical probability as a tool by which to define utility, and then preference, further demonstrates a keen sensitivity toward avoiding the problems which the "privacy" of subjective probability involves in ordinary contexts. Thus, his work is totally within a philosophical analysis of ordinary discourse. Moreover, in following Carnap's mode of analyzing probability, Bohnert adopts a formalism which relates to propositions generally, during spans of time. Thus, as he also observes, this view is not limited to mutually exclusive acts or strategies, as one finds in game theory, but rather deals with the utility of and preference for "any proposition not contradicted by our experience. ..."[55] In pursuing this direction, he finds that he can by-pass the need to consider the complementary aspect of acts and strategies, and concentrate solely upon *what* the proposition involves. Hence, there is greater flexibility in Bohnert's approach, which in itself constitutes an important innovation.

Nonetheless, some difficulties can be pointed out in Bohnert's proposal. However, most of these seem to result from the brevity of his presentation, rather than from any serious flaw in his insight. For example, his definition of "proximate utility" is rendered in terms of the utility of proposition p for x at

---

[55] ibid. p. 229.

time s being "compatible with" the total experience of x at t, together with the estimate of the subject's happiness during time span s beginning at t, with respect to proposition p, in relation to x's total experience at t equaling u utiles.

One difficulty here seems to be with the meaning of "compatible." What sense does it make to say that a proposition is compatible with the subject's total experience relative to a proposition, and the estimate of happiness derived from having p be the case, whatever p may be? Perhaps the suggestion is that in some way the total experience, together with the estimate of happiness, can be sententially expressed, and thus they both emerge as a proposition which is compatible with p. Still, the notion of "compatible" seems in need of further exegesis. In what sense can it be said that two propositions are compatible? His introducing the modal operator for "possibility" does not help answer this question in a definitive manner. Does one presuppose here identity of reference, or truth functional equivalence? Essentially, the same difficulty is manifested with his definition of preference, where Bohnert maintains that proposition p is preferable to proposition q, meaning that the conjunction of p and q "holds" with the utility of p being greater than that of q. As with the term "compatible," *holding* here seems to be in need of further clarification. Moreover, Bohnert would avoid any subjectivistic interpretation of the experience of happiness within his analysis. Evidently what he has in mind is somehow the testable determination of happiness for the subject, as something publicly observable. Furthermore, his allusion to the subject's "total" experience at t is not as nebulous as a first reading may suggest. He goes on to say that it is the total experience at the moment t which becomes the object of the analysis, and not all the experiences in the subject's life. Nonetheless, a case can be made for saying that the individual's *total experience* even at one moment is a vast, almost unimaginable, collection of information.

Apart from the limitations, Bohnert's work is particularly significant in the way it foreshadows an approach to the logic of preference which is based on a logic of events, and thus comes to be a forerunner to Richard M. Martin's way of handling preferences in the latter's *Intension and Decision* (1963). Part of the innovation in what Bohnert does is to allude to preference in an extensionalized event, involving a subject, a relation of utility, a proposition expressing what is preferred, and a time-span, — all extensionally defined. Thus, it can be said that Bohnert extends Carnap's semantics into pragmatics. Carnap himself foresaw the need to investigate this higher level of linguistic

---

[56] Martin, Richard M., *Events, Reference and Logical Form*, The Catholic University of America Press, 1978, pp. 160-161.

analysis; however, he never undertook its systematic study.[56] Interestingly, one observes in the latter part of Bohnert's essay the attention given to preferring as a communication event, such as it is found to be in ordinary discourse. Bohnert's realization — that the only practical way of approaching the analysis of preference is by considering the role agent, the sententially expressed proposition of what is preferred, and a span of time -- is new ground. For it attempts to set down in clear terms the extension of the state of affairs or the "situation" of a sentence expressing a preference.

The significance of the direction chosen by Bohnert, in extending Carnap's semantic theory to a level of pragmatics with the introduction of logical probability, is lost on the editor of the volume in which Bohnert's work appears. In his introduction, Robert L. Davis expresses doubt as to whether Carnap's language system can be extended in this way, which introduces a notion of event so radically different than any encountered in frequency theories.[57] Subsequent developments in the realm of logical analysis were to prove that a systematized pragmatics, as an extension of Carnap's formal language, was indeed feasible.

Bohnert's work constitutes the signaling of an important change in the direction of investigations dealing with the formalization of preference. Bohnert's use of logical probability, which he is quick to disassociate from the remote privacy of subjective probability (being that the former involves in an exclusive sense the reference of a sentential expression) is important in its being a recognition of the need to articulate the logic of preference within a context of language. The importance of this thrust can only be appreciated if the historical development evolved thus far is considered. Not only has Bohnert shown the limitations of the von Neumann-Morgenstern theory in relation to the philosophical analysis of preference, but he offers an alternative which is sensitive to the role of natural language in such investigations. Unfortunately, Bohnert's insight on this point did not receive appropriate consideration in its time. Even Davidson, McKinsey and Suppes, who allude to Bohnert's paper, hardly offer more than passing mention.

With Ramsey's pioneering efforts and the collective attempt by Davidson and his associates, one has the launching into contemporary analytical thought of the logico-philosophical study of preference. However, apart from the vision in these endeavors, the flaw which underlies what they propose is a reflection of the lack of attention given to the workings of natural language itself. This was seen at many points throughout this investigation. For example,

---

[57] Bohnert, Herbert G., "The Logical Structure of the Utility Concept," p. 14.

in von Wright's criticism of Ramsey, one finds that the latter is insensitive to the "accumulative" aspect of the benefit derived from a fruitful course of action. Here von Wright is suggesting that the act of preferring, as a distinct linguistic activity, involves the idea that one chooses to act in a special manner because of a long-term effect which he judges to be conducive to his well-being. Also, von Wright's point that Ramsey's second definition is circular is again derived from the observation of language, where to say that one chooses one course of action as opposed to another already presupposes the operation of the kind of subjective probability which is one's object of definition.

Similarly, one finds in Davidson's paper a recurring expression on the inadequacy of the characterization of preference within the von Neumann-Morgenstern utility theory. Whether reservations are expressed by the authors themselves or by Herbert G. Bohnert, the falling short always seems to have the same *raison d' être*: namely, there is always a *façon de parler* where preference seems to involve the role of chance, or the event of preferring has a unique specificity which identifies it as a particular linguistic phenomenon. It is this which utility theories are unable to grasp.

Within the perspective of their times, these exploratory efforts constitute a reflection of the state of the art. This is to say that one must consider Ramsey's work, as well as the collective effort by Davidson and his colleagues, as indicative of a period in the analytical tradition where a consciousness of language being a "form of life" is simply not present. Ordinary discourse is admittedly not considered a legitimate source of philosophical clarification because of its proclivity toward ambiguity. Moreover, logical investigations, as a method of analysis, is construed in a way where these authors endeavor to capture the nature of preference within a strictly logical framework, without heading any ordinary language context. Logical consistency is seen as the essence of preference-ranking, quite apart from the grammar of preference, as it is manifested in ordinary discourse. The idea which will gain wide currency later on — that logic is a tool which is used to analyze language but not something which can be used to replace natural language — has yet to be realized.

Again, one can allude to the influence of Positivism at this point to picture the climate which may have contributed to the handling of this issue in the particular way in which it appears. Surely, the emphasis Ramsey places on the externalization of partial belief in terms of possible modes of observable action suggests the kind of requirement demanded from philosophical thinking by the Vienna Circle. Along the same lines one can explain the emphasis Davidson, McKinsey and Suppes place on the logical consistency of preference-rankings, as a way of extensionalizing the rational choice underlying preferences, so that it would be free from the inaccessible and empirically unverifiable class of statement dealing with subjective states. While avoiding the temptation to label any of these writers positivists, the standards for philosophical analysis à la the verificationists' school are at least respected by the authors considered above.

Progress in the analysis of preference will be made as thinkers begin to reassess the methods they have used to philosophize in general, where the use of logic as an analytical tool has been reconsidered and where the role of language as an important source of philosophical investigation has gained a respected place.

Chapter 2.

Aristotelean Reflections in Richard M. Martin's Extensionalized Pragmatics of Preference

In his pioneering work on the logic of preference, "Semantic Foundations of the Logic of Preference," Nicholas Rescher refers to Aristotle's *Topics*, Book III, as the first work in which an analysis of preference is presented.[1] The recognition given to the founder of logic is tempered by the observation that though Aristotle speaks of preferability: αἱρετώτερον, his (Aristotle's) remarks are not "formal", in that there is a concentration only upon substantive examples of what is the better or the worse to choose, and that Aristotle does not adequately distinguish "material" from "formal" considerations.

Rescher's reference to Aristotle does not go beyond being a historical homage, and he underscores the need to develop preference-principles which are acceptable "upon abstract, formal, (and) systematic grounds."

Rescher's remarks will be found to fall short of seeing the depth with which Aristotle endeavors to evolve preference-principles, which have a remarkably modern ring. In fact, where Aristotle's work in *Topics* III is compared to Martin's work on preference in *Intension and Decision*, the former's astonishing insight into certain elementary preference relationships is unmistakable.[2] Martin's work is chosen so as to introduce the things Aristotle says in this area partly because there is a pronounced similarity between Aristotle and Martin respectively, in their approach to the analysis of preference. This in turn provides a means of not only introducing the Aristotlean contribution, but of also seeing in Martin's more thorough treatment of the subject a philosophical perspective which has its origins in a tradition which goes back further than is sometimes realized.

In part, the aim of this chapter is to explore the extent to which it can be said that Aristotle is proposing the seminal beginnings of a logic of preference

---
[1] Rescher, Nicholas, *The Logic of Decision and Action*, University of Pittsburgh Press, (1966), p. 38.
[2] Martin, Richard M,. *Intension and Decision*, Prentice-Hall, (1963).

in the *Topics*, III, as well as in the *Rhetoric*, Book I. For contrary to Rescher, it seems that more than just "substantive examples" are at work in Aristotle's account. Moreover, the steps Aristotle is taking have a great deal to gain in clarity when seen in light of R.M. Martin's efforts in analyzing preferences within a formalized extensional pragmatics. The precise point of tangence between what one finds in Aristotle and in the direction of Martin's treatment is that both are keenly sensitive to the characterization of preference in a manner which brings forth the consequences of practical choice by a responsible agent. Significantly, the axioms Martin specifies as underlying preferential decision making can be said to be foreshadowed in Aristotle, in terms of their being basic principles governing a choice of "correct" actions within the context of a variety of practical concerns pertaining to dialectic.

If only from the viewpoint of the history of ideas, it is fruitful to investigate Martin's work not only in itself, but also in light of its intriguing Aristotlean reflections. More no doubt comes into focus by considering Martin's contribution in this way, and one has the added benefit of understanding the importance of Aristotle's real discoveries in this area, which have not been adequately discussed by commentators.

On the surface, Martin's efforts seem to fall along a somewhat similar line of investigation as that encountered in Herbert G. Bohnert's work in 1954. This is to say that both Bohnert and Martin approach the study of preferences from the standpoint of specifying the nature of the "event" which is involved in preference. Bohnert, it will be recalled, attempted to account for the kind of event which is involved in preferences by seeing it as a probability function. His ultimate rendition of preference is in terms of what he calls "proximate utility", which is characterized as an event involving an agent, x, some proposition expressing a possible course of action, p, a time span, s, the agent's happiness, H, etc. Thus, Bohnert attempts to identify and include in his characterization of a preferential event all those pertinent elements which are conceivable parts of the probability-event of x preferring one course of action as opposed to another. In its time, Bohnert's contribution manifested a new sensitivity toward explaining what "it means to say" that one prefers a to b. However, it cannot be said that Bohnert was approaching the study of preference purely from the standpoint of characterizing preference as manifested in natural discourse. This is to say that Bohnert was still much influenced by the von Neumann-Morgenstern approach, and he attempted to avoid some of its vagaries in philosophical circles by appealing to the Carnapian notion of *probability*$_1$, thereby hoping to explain more fully the preferential state.

Martin, on the other hand, though he is equally concerned with the correct characterization of the kind of event which preference involves, proceeds in a manner which more closely follows the sense of discourse concerning preferences, as manifested in an experimental framework. His approach is to make explicit the extension of discourse involving preferences, taking into account an entire context of scientific testing. Thus, the "preference-event" is expressed in terms of a language user, in relation to preference as a sententially expressed entity, extensionally considered in terms of preferred object, span of time, etc. Thus, without entering into additional detail at this point, one can appreciate the definite innovation Martin's approach is introducing in the area of prohairetic logic, through his conviction that certain forms of discourse, under specific and prescribed conditions, can supply the elements for indicating the logical form of preferences, and of their hierarchic structure. This again, that is the reference to specific situations and conditions of speech with regard to things, is the intriguing aspect of his work which brings him in proximity to Aristotle. For one finds in the latter a similar sensitivity for explaining what it *means to say* that one prefers a particular course of action as opposed to some other. As will be seen, Aristotle is also very much committed to determining actual cases experimentally when attempting to explain what one means by preferring one course of action as opposed to another.

Though Martin's work in this area is more within the trend of examining the constituents of discourse by means of the simplest logical tools, so as to devise a tentative formalization of preferences, this should not be regarded as the advocacy on his part of an appeal to ordinary discourse as a source for discovering the fine points of a logic of preference. In essence, he would not be above agreement with Davidson, McKinsey and Suppes that ordinary discourse is fraught with ambiguity and shifting nuance, which makes it unsuited as a source of philosophical knowledge. It is worth reminding oneself that Martin's allusion to discourse here is to discourse under *controlled* experimental conditions. Moreover, he presupposes a clearly defined communication event between examiner and respondent. This is not an appeal to "ordinary discourse" in the sense of an open-textured, ever-altering linguistic form of life. Though he eschews reference to natural discourse in this sense he is still keenly aware of the fact that at best one can only have the formalization of preferences, and of their rankings, which is tailor-made, so to speak, to the individual respondent who is being tested to discover his choices and subsequent preferences. Thus, the variability which Martin sees in the formalization of preferences derives itself largely from the idiosyncrasy of the individual's choices, and is expressed by a very skillful handling of logical technique. However, it is not a variability which is reflective of the changes and shifts in meaning which natural discourse is subject to.

Regarding the first phase of the investigation into Martin's logicized semantics of preference, it is important to orientate oneself to the fact that N.Rescher, in denying that Aristotle had anything to do with preference principles in the *Topics*, may be said not to have perceived the purpose for which Aristotle wrote this treatise. Quite simply, the subject matter of the *Topics* in Book III is to define and discuss the art of dialectic. In various ways, Aristotle is attempting to demonstrate how the best or more plausible position in debate can be recognized and assessed. His aim in the *Topics* III is removed from any disposition to say something concerning ideal forms, ontic commitment, or about the proper schemas of syllogistic reasoning. Aristotle simply presents a series of rules which he believes serve as guidelines for successful disputation. In doing this he offers no "philosophical" view of his own. Rather, his intention is to see the "practical" consequences of choice between related states of affairs. Thus, he appeals to common sense for support of his observations that in all rational disputation there are certain general principles which persons of good will must respect.

Unfortunately, Rescher sees only the content of the examples Aristotle employs, and infers that all he is doing here is setting down non-formal, i.e. substantive, principles. He does not allow for the possibility that perhaps Aristotle may have been looking at the act of disputation meta-linguistically, and that perhaps he was endeavoring to discover rules which went beyond the parameter of the examples which he was bringing forth to illustrate his insight. One can hardly accept the view that the scope of Aristotle's presentation was designed to stop at the narrow confines of the cases which he discusses, especially in the context of the remarks he makes concerning the broader applications of his investigation in Book III.

The working hypothesis upon which this chapter is based is that indeed Aristotle had arrived at an understanding of principles which guided human choice in successful disputation. Insofar as he claimed that rational discussion demanded the adoption of some rules, it is not far of the mark to conclude that he sees these rules as constituting the semblance of a logic of preference. Though as a term "logic" should not be loosely applied to just any collection of directives, it is nonetheless warranted in this case since the principles Aristotle alights upon are all closely allied to the single goal of achieving success in disputation by the most prudent and therefore rational means.

An objection to what is being proposed may be that in the *Topics* Aristotle did not attempt to articulate a scientific theory regarding truths, but rather sought to set forth rules for a non-scientific account concerning plausible discussion. Though it would be wrong to claim that in the *Topics* Aristotle intended to develop precisely what comes out in his *Analytics*, this is no reason to deny that the two techniques involved both in analytics and dialectics are *equally* scientific. As E. Weil correctly points out, analytics deal with the formal character of scientific reasoning, dialectics deal with the "formal character" of reasoning regarding discussion, regardless of whether or not in the latter case the discussion is about propositions whose truth is only plausible.[3] Both in the analytics and in the dialectics one is dealing with rules of an equally formal character. The subject matter of each has nothing to do with specifying the *validity* of the methods involved in both cases.

It would also be a mistake to think that because Aristotle in this book of the *Topics* spoke mostly in terms of concrete examples of what is "better", that the rules he presents are so context-dependent that they are not intended to have practical use beyond the immediate cases he mentions. Toward the close of Book III, Aristotle makes it a point of mentioning that all the rules he has brought forth can be considered in a more "universal" way by altering their mode of expression.[4] Thus, he does not see what he has done as merely substantive and not formal. Basically, it is the demonstration of the formal facet of Aristotle's rules of dialectic in Book III which is also in part the object of this chapter.

Prior to embarking upon the demonstration of these preference-principles, which though found in Aristotle have a surprisingly modern ring, it is important to make clear the sense in which Aristotle uses the word αἱρετώτερον to express the idea of the preferential. The expression αἱρετώτερον is loosely translated in the English texts as the "desirable". However, its meaning should be understood more accurately in terms of that which is instrumental in securing some goal or end. There is scarcely anything in the primary sense of this word which suggests a subjective state of consciousness or dispositional attitude, such as "want" or "desire".[5] This is a vital point since it serves to clarify again the highly practical context in which Aristotle sees

---

[3] Barnes, J., Schofield, M., Sorabji,R., *Articles on Aristotle*, G.Duckworth & Company Ltd., (1975), pp. 89-90.

[4] Aristotle, *The Works of Aristotle*, W.D.Ross, editor, Volume I. *Topics*, Book III, Oxford University Press, (1968),Chapter 5, p. 119a.

[5] Liddell and Scott, *A Lexicon*, Abridged, Oxford, (1963), p. 21.

preferability. In a sense, it could be said that the relations of preference are considered by him in an extensional and pragmatic way, meaning that preferability is seen in terms of the practical, and, therefore, observable effects one *kind* of spoken disputation has as opposed to another.

In this connection it is important to keep in mind that the determination of how one opinion is found to be "better" than another is gotten by Aristotle through the art of testing (*peirastikè*). The latter considers what the "wise" or the knowledgeable would *accept* as the better of two opinions. Some essential ideas are seen to emerge here relative to Aristotle's treatment of the rules of preference. First is the general notion that the preferable results from testable determinations. Secondly, one has the more specific observation that the determination of a preference results from observing the accumulative acceptability of some verbally expressed state of affairs. In a sense therefore, Aristotle extensionalizes the idea of preference both where he introduces the process of testing and the observation of public acceptance. Though he grounds the basis of preference upon a kind of knowledge which Aristotle terms "ένδοξα", this should not be taken in the pejorative light in which Plato comes to talk of knowledge by "δόξα" in the *Republic*. Rather, "ένδοξον" should be taken as the state of *informed* knowledge, and not knowledge in some inferior sense as was the case with Plato's conception of δόξα. Again Weil's views on the matter are highly informative. He points out that Aristotle's rules are ways of acquiring victory in inquiry.[6] As such, they are a heuristic means for achieving correct premises for syllogisms. Thus, these rules serve as a way of testing the immediate strength or weakness of knowledge concerning a limitless variety of things.

An important point emerges at this juncture relative to the special status of the subject matter dealt with in the third book of the *Topics*. For in the context of the historical development preceding him, it would have been very important for Aristotle to base the art of responsible disputation upon a rational footing, one which did not lend itself to the sophistic attitude of *Protagoras* or of *Gorgias*; where the disputant was concerned with the "appearance of truth," or of how to make the worse *appear* the better cause. It is interesting to observe how an outline of a logic of preference appears in Aristotle within the framework of attempting to discover correct standards for disputation, i.e. dialectic. That which is preferable is so because it is that which has a rational basis of acceptance in discussion by reasonable persons. Here one can see something

---

[6] Barnes, J., Schofield, M., Sorabji, R., *Articles on Aristotle*, pp. 92-93.

of the conviction expressed earlier by Socrates and Plato, namely that discussion by well-intentioned persons must never involve deception as its goal, but must exhibit rational decision-making toward a determination of the best view with respect to the subject at hand. Thus, the logic of the preferable is highly significant in that in its way it serves to differentiate the sophistic from the philosophic attitude toward truth. This specific point is faced more directly by Aristotle in a work closely allied to the *Topics*, namely the *De Sophisticis Elenchis*. There he alludes to the way in which the sophists attempted to teach not by stressing the importance of reason in argumentation, but in giving their students prepared speeches to recite.[7] Primarily, the sophists were concerned with the appearance of wisdom, as opposed to the training of the mind to reason concerning generally accepted opinions, and thus arriving at conclusions which were defensibly on the art of reasoning, rather than upon public spectacle.

Thus, the first objective of the present investigation is to demonstrate in broad outline how one can recognize in Aristotle's *Topics* III certain vitally important preference-principles, together with their underlying conceptual justification, which analysts have come to regard mistakenly as insights first realized in this century. The point of all this is not to idly embellish Aristotle's prominence as a founder of dialectic, but hopefully to redirect some attention to this particular aspect of the *Topics*, and perhaps to see in it more than has been realized before on the theme of the development of the logic of preference.

As a means of accomplishing the above, germane portions from Richard M. Martin's work, *Intensions and Decisions*, will be used to show the close parallelisms between what is found in Aristotle and certain elementary aspects of Martin's contributions. The remarkable closeness in the way both of these thinkers attempt an axiomatic approach to the analysis of preference and choice cannot but bring home the relevance of Aristotle's work to contemporary logical analysis, at least in this specific area. For in some cases, Martin's instructive insights are found to be first gleaned by Aristotle.

As noted, Martin investigates the semantics of preference by considering the kind of formal relationships which can be brought forth between a language user and the extension of a sentence which conveys what is preferred. However, prior to demonstrating how some of these relationships are manifested in Aristotle, it is striking to observe how even in the introductory re-

---

[7] Aristotle, *The Basic Works of Aristotle*, Richard McKeon, editor Random House, New York, (1941), *De Sophisticis Elenchis*, 183, 35-40.

marks Martin presents to his formalization of preferences, there are mirrored Aristotle's preconceptions concerning the nature of choice and preference.

In a central passage Martin says:[8]

> "Suppose E presents the subject X with two sentences a and b of the language L at time t. And suppose he *asks* X which of the sentences X regards as the *more probable*, or in which he *more firmly believes*, or whether he is more willing to *accept* one than the other, or whether he is more willing to base *his actions* on one than on the other. E then himself accepts and perhaps records a sentence of the meta-language to the effect that X at time t indicates preference for sentence a over sentence b, if such is the case. Strictly E gives X the *choice* of the two sentences and records which of the sentences X has chosen. Several hours later, or the next day or month, E might repeat the experiment with X and get the same choice in response. E would reasonably infer, there being no evidence to the contrary, that X *would have made* the same choice at any time in between. In other words, E concludes that X prefers a to b during some time interval, he is then presumedly willing to infer that X would choose a to b at any time during that interval. Preference then may perhaps be handled as continuous choice (whether explicitly verbalized and recorded or not) throughout some time-interval...."

One notes here how Martin sees preference in terms of that which is tested and observed to be the choice of the respondent X. Furthermore, choice, which is preference over a short time-interval, is rendered in terms of that which is acceptable to the person tested. The entire conceptualization of preference is rendered in terms which involve the experimental determination of patterns of observable conduct over a span of time.

Turning to Aristotle's introductory remarks in Book III, one finds virtually similar concerns expressed. With the substitution of the word "preferable" for the word "desirable," W. A. Packard-Cambridge's translation reads as follows:[9]

> "The question: which is the more (preferable), or the better, of two or more things, should be examined upon the following lines: only first of all it must be clearly laid down that the inquiry we are mak-

---

[8] Martin, Richard M., *Intension and Decision*, pp. 45-46.
[9] Aristotle, *The Works of Aristotle*, W.D.Ross, *Topics*, Book III, 116a, ll. 1 to 14.

ing concerns not things that are widely divergent and that exhibit great differences from one another (for nobody raises any doubt whether happiness or wealth is more preferable), but things that are nearly related and about which we commonly discuss for which of the two we ought rather to vote, because we do not see any advantage on either side as compared with the other. Clearly, then, in such cases if we can show a single advantage, or more than one, our judgment will record our assent that whichever side happens to have the advantage is the more (preferable)."

Above, Aristotle sees the determination of what is preferable in light of considering two or more "closely" related things, concerning which people express some kind of acceptance or vote. Thus, preference results from the analysis of an *observable statement* of choice. This is not far removed from R. M. Martin's requirement that preference be understood in terms of an experimental procedure which interprets a subject's spoken expression of acceptance. Moreover, Aristotle's perception that preference deals with things which are "closely" related reveals an understanding of the fact that the preferential is not an *obvious* determination, but results from a careful scrutiny of similar kinds of things. One can point to an implication of this view of preference, namely that time plays an important role in allowing for the gradual recognition of the preferable. Remarkably, Martin's realization of how preference is a determination one makes after the examination of choices *over a period of time* almost precisely represents Aristotle's observation that preference is a judgment one makes "after" a determination has been made as to which side (of an argument) is chosen as more advantageous. Furthermore, the relationship of Aristotle's use of the word αἱρετώτερον for preference to the verb αἱρέωμαι (to choose) demonstrates again a similar kind of logic operating in Aristotle's thinking as in Martin's presentation.[10]

It is important to reiterate the point brought out in Aristotle that the preferable deals with *closely* related things. For preference must represent a "fine" distinction between possible options. It makes little sense to say that one has a preference between two unrelated things. For example, to prefer tea to skiing makes no sense unless there is a context which specifies a goal for which these two options are rationally connected. Aristotle recognized this when he speaks of things which are "closely" related becoming the subject matter of that which is preferable. The discussion which leads to the determination of

---

[10] Liddell and Scott, p. 21. ll. Aristotle, *The Work of Aristotle*, W.D.Ross, *Topics*, III, p. 117.

what is preferable is actually the testing stage. In Martin's rendition, the determination of the preferable is also given in terms of experimentation and testing. The latter implies that the preferable is a "fine" distinction between two or more related states, ranging over a measurable period of time. It is entirely proper that Martin alludes to the experimental method as a means of explaining just how preferences are determined. For it is under experimental conditions that what seem to be *plausible* options are tested for their acceptability by the experimenter. Here again the idea of determining which of the *closely* related options constitutes the subject's preference emerges.

Proceeding on to the consideration of preference-principles in Aristotle, the strategy will be to show how many of Martin's insights into the formal relations involving preferences can be recognized in Aristotle's work. So as to make the comparison more palatable, certain adjustments will be made in the interpretation of Aristotle's text. For the precision found in Martin's work could scarcely be expected in the *Topics*. However, once it is realized that Aristotle is concerned with the useful or practical context of dialectics, from which he makes his observations as to the nature of the preferable, then their differences are not found to be unbridgeable.

One of the first points of comparison is with respect to Martin's innovative method of considering the relations of preference in terms of relations between a language user and the extension of his uttered preference, considered sententially. If it can be demonstrated that Aristotle's approach is somewhat similar to Martin's in this respect, then a further bond can be forged between these two thinkers, which in turn will facilitate the comparison being proposed. Interestingly, in his translation of the *Topics*, Packard-Cambridge introduces the symbols 'A' and 'B' to explain the relationships between things which are preferred, and how all the constituents of preferences are said to be each individually preferable in relation to the constituents of things which are not preferred. Hence, on a surface reading, it may appear that 'A' and 'B' here may be taken as designating sentences or sentential expressions of some sort, and that this would be a possible link with the thrust of Martin's analysis. However, preceding and following the present passage Aristotle refers to the case of a man being better than a horse, and that the best of men is better than (more preferable to) the best of horses. Thus, it is quite clear from the deeper sense of the passage that Aristotle is not referring to sentences being preferred as say more truthful than other sentences, and that the translator employs 'A' and 'B' to designate things. It is important to note, moreover, that in the original Aristotle does not use any symbolism in this passage.[11]

---

[11] 32-35.

Yet Aristotle goes on to *generalize* in this very same passage, saying that "Man" is preferable to "Horse", since any particular man is preferable to any particular horse. In view of this, he is applying his insights into preferences not only to things, but also to classes of things, e.g. Man, Horse, etc. Thus, the issue becomes whether the character of the analysis Aristotle offers has been changed with his introducing the idea of preferences between classes of things.

Though it cannot be maintained explicitly that Aristotle develops his analysis of preference in terms of a fully articulated extensional pragmatics, a number of factors come into play in the *Topics* to suggest that in some respects what he has to say about the formal aspects of preference parallels the pragmatic analysis Martin develops in a more systematized fashion. Indications of this are found in the fact that (1) the subject matter of the *Topics* is devoted to the art of dialectics, or the rules for successful *spoken* disputation, (2) that these rules precipitate from experimentation with and testing of *uttered* responses, (3) the rules of preference apply not only to cases where specific things are preferred, but in a more all inclusive sense, they apply to cases where one prefers a class of things over another class of things, and (4), as noted already, Aristotle states toward the end of Book III that his rules of preference can be interpreted in more general terms, which need not reflect any reference to specifically particular things. These factors support the view that Aristotle is considering in a suggestive manner the formal facets of preferences, within the sphere of a communication event, and that his vision is not limited to the specific cases he employs to illustrate his observations. Thus, in essence, it can be claimed that in the context of developing a dialectic, Aristotle has recognized some formal aspects of preference with an awareness of the role of language, which in a general way parallels the pragmatic approach of observing linguistic use in relation to a language user, and the acceptance or rejection of some specific sentential expression.

Consequently, it can be argued that where Aristotle speaks of the irreflexibility, asymmetry, and transitivity of some of the relations of preference, he can be interpreted as meaning that a linguistic expression of some assertive force expresses the content of the particular preference involved. Fundamentally, however, one still cannot claim that Aristotle sees preference relations in terms of an individual "uttering" a preference with respect to a clearly identifiable sentential expression, extensionally specified. This is R.M. Martin's unique innovation in the area. However, neither can it be denied that Aristotle recognized the possibility of a logic of preference within a sphere of spoken discourse. This discovery is of great importance for Aristotle since it enables him in *De Sophisticis Elenchis* to propose a program for dissolving sophistical arguments, as indicated earlier. Hence, the common ground between the two approaches being considered seems to be that one can interpret Aristotle as devising a logic of preference which *could* lend itself to involve sententially expressed preferences. Just as, on the other hand, Martin proposes that his observations concerning preferences and their rankings could be seen differently. This is to say that Martin claims that in certain cases the same kinds of relationships of preference are found to hold between goods or things as between sentences about such items.[12] Furthermore, the element of temporal sequence, which is implicit in Aristotle's principles, shall be made explicit for purposes of this comparison. This is a means by which to bring Aristotle's text closer to Martin's, in that the latter uses variables to range over spans of time so as to effect a sharp precision as to the unique series of events which are manifested in complex preferential relationships.

First, however, under "General Rules of Preference" Martin sets down the following rule: PR1. XPrfr $a, b,$ t $\supset$ (Sent $a \cdot$ Sent $b \cdot - (a = b)$).[13] The above says that if X prefers a to b, then a and b are distinct sentences of L.

PR1. actually specifies the irreflexibility of the relation expressed by 'Prfr'. It describes the exclusivity of X's preferring a, in relation to b. Preference exhibits the particular trait, namely that it is always seen in light of at least two possible options, and that the *acceptable* option is always found to be different from the *nonacceptable* one because of its being chosen by some agent. This comes out quite sharply in the implication expressed in PR1. For sentence a is not identical to sentence b, if sentence a is preferred to sentence b. Hence, one cannot but admire the clarity with which Martin expresses the idea that since a is preferred to b, it follows that the negation of the identity of a to b is the case. Thus, the fact that a is preferred to b, differentiates a completely from b, within the context of preference.

Turning to Aristotle, one finds virtually the same rule expressed at the beginning of Book III. Returning to the passage already quoted one's attention is held by the following line:[14]

> "...Clearly, then, in such cases if we can show a *single* (my italics) advantage, or more than one, our judgment will record our assent that whichever side happens to have the advantage is the more (preferable)."

---

[12] Martin, Richard M., *Intension and Decision*, p. 59-60.
[13] ibid., p. 50.
[14] Aristotle, *The Works of Aristotle*, p. 116a, ll. 10 to 14.

Aristotle is making the preferable dependent on the advantageous, even if that advantageous aspect may involve a single advantage. Keeping in mind that for Aristotle the advantageous deals with that which is found to be instrumental, the basic idea expressed by Martin is being suggested by Aristotle. For a number of things may be instrumental in securing the same goal. Yet where the goal can be secured by a means which is found to be even slightly better than another, then that means will be different from all the others *since* it will be preferred. Hence, to prefer makes that which is preferred different from any other thing which may be "closely" related to it. In essence, one sees here the basic elements of preference: the reference to at least two options, and the way in which the preferred achieves a status separate from those things it is usually related to.

Martin's second "general" principle states:[15]

PR2. $X \operatorname{Prfr} a, b, t \equiv (t_1) ((\operatorname{Mom} t_1 \cdot t_1 Pt) \supset X \operatorname{Prfr} a, b, t)$. The meaning of this rule is that where X prefers a to b at time t, this is equivalent to saying that for every momentary time-span t_1, which is part of time t, then X prefers a to b at t_1. Martin states that PR2 makes explicit the role of choice in the broader concept of preference. For the choice at t_1 is a momentary part of X's preference.

The counterpart of PR2 in Aristotle first occurs after his remarks concerning the close relatedness of options of preference. It is interesting to note how both writers see the same sequence between a preference principle which deals with the exclusivity of what is preferred, and a preference principle which specifies the elemental constituents of preference. In Aristotle, the latter occurs where he says that the preferable is that which is most likely chosen by the prudent or knowledgeable. The suggestion here is that what is consistently accepted by the experts in some particular area will become the preferable for that domain of interest. In essence, consistent choices form the concept of preference.

However, a more precise representation of the idea expressed by PR2 occurs in the following passage:[16]

"..., if A be without qualification better than (preferable to) B, then also the best of the members of A is better than the best of the members of B; ...Also, if the best in A be better than the best in B, then also A is better than (preferable to) B without qualifications; ..."

---

[15] Martin, Richard M., *Intension and Decision*, pp. 50-51.
[16] Aristotle, *The Works of Aristotle*, p. 117b, ll. 32-35.

The above passage brings out how that which is preferred is such that any constituent part of it is preferred to any counterpart of what is not preferred. Remembering that for Aristotle the constituents of preference are choices, it is uncanny that he insists upon an internal uniformity to preferences so that any local part of the preference will exhibit the same advantageous character as the whole preference, as a totality. The anticipation of the kind of insight Martin explicitly brings out is remarkable, especially because in the above passage Aristotle goes on to say that if a segment of a preference is said to be more advantageous then a segment of what is not preferred, then one can infer that the total collection of segments in the former case constitutes a preference in relation to the total collection of segments in the latter case.

Aristotle's articulation of preference principles can also be exemplified by showing how he has arrived at a notion of "rational preference ranking." Martin introduces the latter idea in terms of the notion of indifference. He defines the relation of indifference as "equality of preferential status." What is meant here is that one would be indifferent to sentence a or sentence b if they were both preferred in some identical situation. Martin's formal definition of "rational preference ranking", RPR, relative to X and t runs as follows:[17]

'FRPRX t' abbreviates '$((Ea)(Eb)(Ec)$ $(Fa \cdot Fb \cdot Fc \cdot \cdot -(a = b) \cdot \cdot -(b = c)) \cdot (a)(b)(c)((Fa \cdot Fb \cdot Fc \cdot X\text{Prfr } b,c,t \cdot \cdot -(a = c) \supset X\text{Prfr } a,c,t) \cdot (a)(b)(c) ((Fa \cdot Fb \cdot Fc \cdot X\text{Indiff } a,b,t \cdot X\text{Indiff } b,c,t) \supset X\text{Indiff } a,c,t) \cdot (a)(b)((Fa \cdot Fb) \supset (X \text{ Prfr } a,b,t \text{ v } X \text{ Prfr } b,a, t \text{ v } X \text{ Indiff } a,b,t) \cdot (a)(b) ((Fa \cdot Fb \cdot X \text{ Prfr } a,b,t) \supset -X \text{ Prfr } b,a,t))'$.

Martin's definition takes into account three distinct sentences a,b, and c, such that if a is preferred to b, and b is preferred to c at some time t, then a is preferred to c. Moreover, where X is indifferent to a over b, and b over c, then X is indifferent to a over c, at time t. And where a is preferred by X over b, or b is preferred to a, or X is indifferent to a over b, then X does not prefer b to a.

The restrictions Martin builds into his conception of preference rankings are meant to reflect possible specific instances where preferences may be involved. However, he makes a point in insisting that preference and indifference do not always exhibit a "quasi-transitivity." Martin is acutely aware of the fact that there simply are situations where one cannot always infer a preference from antecedent conditions.

Though again the precision of Martin's text could not possibly be expected in Aristotle's, enough of the seminal ideas of a rational preference rank-

---

[17] Martin, Richard M., *Intension and Decision*, p. 53.

ing is found in the *Topics* to suggest that he sees some of the basic issues which this idea involves. However, in this case a further clarification of how Aristotle sees such ranking is also found in the *Rhetoric*, Book I.

In the *Topics* the following statement indicates preference ranking is being entertained by its author:[18]

> "...if there be two things both preferable to something, the one which is the more preferable to it is more (preferable) than the less highly preferable..."

One sees here a number of things which suggest that preferability is a matter of successive levels. For example, Aristotle is seen as speaking in terms of "both things being preferable to a third," which comes very close to saying the kind of thing Martin refers to where the latter defines indifference in terms of "equality of preferential status." Moreover, one finds the operation of distinct levels such as the "more highly preferable" as well as "the less highly preferable." However laconic Aristotle's remarks may be, similar kinds of rankings are observed here as in Martin's text. Essentially, one finds in Aristotle that there is the basic level of the preferable, the level of the indifferent (where two things are equally preferable), the level of the more highly preferable, and the level of the less highly preferable. One also notices how these levels are telescoped, so that one arrives at a notion of indifference through understanding how two things may both be individually preferred relative to a third. Thus, one arrives at a notion of indifference in terms of the more basic notion of preference. Furthermore, higher preference is really based upon the notion of indifference, and the same is true of the notion of lesser preference.

Aristotle alludes to the same sort of preference ranking in the *Rhetoric*, Book I, where he says:[19]

> "....when two things each surpass a third, that which does so by the greater amount is the greater of the two; for it must surpass the greater as well as the less of the other two...."

It scarcely needs repeating how the above demonstrates the recognition of different levels of preferability. More interestingly, however, Aristotle makes it quite clear that the regularities which one sees operating within the different levels of preference are greatly dependent upon the nature of the thing which is being compared and analyzed. Thus, one sees to some extent how Aristotle

---

[18] Aristotle, *The Works of Aristotle*, p. 118a, ll. 118b, 2-4.

[19] Aristotle, *The Basic Works of Aristotle*, Richard McKeon, editor, *Rhetoric*, Book I, p. 1347, ll. 33-35.

is concerned about setting forth rules for the different levels of preference which are not pervasive for all possible situations. This sensitivity to the way that context influences the formal rigor of preference-principles was found to be very clearly expressed above in R.M. Martin's approach to the subject, where he spoke of the quasi-transitivity of relations of preference.

The passage from the *Rhetoric* is significant beyond the point of simply exemplifying Aristotle's insights into the nature of the rankings of preferences. Where one considers his remarks at the opening of this work, one sees that he relates rhetoric to dialectic so that the former is the "counterpart" ('αντίστρο-φοσ) of the latter. By the first, one intends to demonstrate the order for discussing and maintaining the truth of statements, whereas the second deals with persuading an audience of the falsehoods of allegations. The latter is accomplished by the use of enthymatic arguments. Thus, both dialectic and rhetoric deal with the truth of statements. However, the former seeks to ascertain the correct mode of inquiry for attaining true statements, the latter seeks to put statements in their proper perspective by revealing linkage to antecedent and consequent truths. Here again Aristotle makes it a point in differentiating rhetoric from sophistry, which merely deals with the appearance of truth.

Preference operates within rhetoric in that rational beings prefer to be persuaded by that which is fully revealed in the entirety of its ramifications. Thus, as the word 'αντίστροφοσ suggests, rhetoric is the counterpart of dialectic in the sense that rhetoric is *a special application of* dialectic. Of additional importance to the study at hand is that Aristotle makes it quite clear at the opening of the *Rhetoric* that dialectic deals with the truth of statements: "... πάντεσ γαρ μέχρι τινοσ και υπέχειν λόγον. ..."[20] One asks again whether Aristotle conceived of his principles preference in terms of sentential expressions. The matter now seems closer to resolution at least up to the point of saying that in the *Topics* Aristotle is developing a logic of preference in terms of statements expressing choices. Thus, the relationship to Martin's approach comes dramatically closer in view of Aristotle's use of "λόγον" at the outset of the *Rhetoric*.

The fact of Aristotle's awareness of formal preference principles has been demonstrated to some extent. What remains by way of completing the comparison is to consider some of the ramifications of his insights as these pertain to a new understanding of the *Topics*, as well as to the questions which seem necessary as prerequisites to presenting a logic of preference.

---

[20] Aristotle, *Aristotelis Opera*, II, Academia Regia Borussica, Berloini Apud W. de Gruyter et Socios, MCMLX, p. 1354.

Earlier the observation was made that in the *Topics* Aristotle considers language as manifested in disputation "meta-systematically." This point is central since it describes Aristotle's innovative procedure in Book III. As briefly stated in *Aristotle's Concept of Dialectic* by J.D.G. Evans, Aristotle's intent in the *Topics* is to concentrate on the technique which allows advance from qualified intelligibility to unqualified intelligibility.[21] In this, Evans is saying that dialectic (in Book III) deals with showing *how* one can recognize what is seemingly true (qualified truth), which in turn will become a true premise (unqualified truth), and thus serve as a link in a valid syllogistic argument.

What Evans' remarks do not explain is that one cannot set forth principles for what is preferred unless the spoken expression of choice is in some way "bracketed". If this were not the case, then one could not advance beyond the particular case of a preference. Aristotle is seen to reject any such limitation being placed upon his study of the nature of the preferable at the end of the third book. This suggests that the investigation into what is preferable must proceed by considering language in some meta-systematic way, so that the preferred, as an expression, has certain specifiable and formalizable relations to its user, as well as to other statements within a speaker's discourse. When carefully considered, the very ideas relating to precisely how choice plays a part in the emergence of preference relations, the notion of preference ranking, etc. could not be had unless discourse is seen as in some sense *spoken about*, and made an object of observation.

Thus the *Topics*, and especially Book III, is perhaps historically one of the first works where discourse is looked at as an "object" of analysis. In doing so, however, the *Topics* raises an interesting issue, namely if dialectic deals with the analysis of discourse with a view of setting forth formal rules by which the plausible can be discovered, how can these rules exhibit any "rigor" if they are about that which only seems to be true or acceptable? In other words, in attempting to give general rules for preferences, how can one claim that any sort of rule can be given for that which is relatively true or at best only plausible?

Indeed, recent commentators observe that the reason why this particular work by Aristotle has been looked upon as "immature" is because it was considered a bad attempt to arrive at a syllogistic logic, such as one finds in *Analytics*. It was not realized until quite recently that the *Topics* has a specific subject matter, which forms a methodological prerequisite for formal logic as we see it in Aristotle.

---

[21] Evans, J.D.G. *Aristotle's Concept of Dialectic*, Cambridge University Press, London, (1977), pp. 92-93.

The distinction of this subject matter perhaps can best be appreciated as an attempt to evolve a formalization of preference-principles. Yet the difficulty noted above relative to the question of how these rules can exhibit any semblance of rigor when they are found to deal with only what seems plausible in disputation, tends to minimize for critics the importance of the enterprise in the *Topics*, and especially of Book III. For it seems that one is caught up in a paradox when attempting a logic of preferences. On the one hand, one must have a "logic" in the traditional sense of a formal system with axioms which enable one to prove theorems, etc. On the other hand, since preference is itself a relation which is heavily dependent on the context one is working with, it does not seem possible to have a logic with a rigor which is also sensitive to the nuances of context.

Aristotle's approach to this problem is to propose certain general principles as self-evident axioms, which have been manifested by disputation, entered into by individuals who sincerely desire to reach true opinion. Though his axioms do not go much further than to indicate the irreflexivity, asymmetry, and transitivity of the relation of preference, they still represent the most fundamental advance in the field of developing a logic of preference to date. This is to say that the insights brought out above, constitutes a core of preference-principles which most analysts who employ the axiomatic approach in the analysis of preferences agree upon as intuitively acceptable. It is quite incredible that scarcely anything has been said in the analyses of this area about Aristotle's work, and the innovation it is proposing. As noted at the outset, Rescher's brief reference to the *Topics* hardly considers the significance of Aristotle's remarks, since the former does not distinguish the formal from the substantive aspect in the latter's views.

The issue of how one is to determine precision within such a "logic" is handled by Aristotle in a way which demonstrates his sensitivity to different criteria for various kinds of systematization. For his analysis of disputation in Book III is predicated upon observing the discourse of persons expert or knowledgeable in some particular area of science:[22]

> "First, then, that which is more lasting or secure is more (preferable) than that which is less so: and so is that which is more likely to be chosen by the prudent or by the good man or by the right law, or by men who are good in any particular line,...."

---

[22] Aristotle, *The Works of Aristotle*, W.D.Ross, editor, Volume I, *Topics*, p. 116a, ll. 12-16.

Hence, where one has a community of individuals which is interested in attaining right opinion about something, then there are certain predictable relationships between those statements which this community would unreservedly choose as acceptable under certain conditions, and those which would not be so chosen. Thus, the consistency of general preference-principles must be considered in the context of earnest inquiry. Once the latter is had, than these principles are found to be operative in relationally ordering the progress of discourse in any given area. Though preference-principles are derived from *observing* disputation, they achieve a measure of generality or quasi-universality because they constitute the basic ground rules of productive discourse generally. The *Rhetoric* provides an insight into the kind of generality which Aristotle intends these principles to have in the *Topics*. In a revealing passage he says: "...καθάπερ και 'εν τοῖσ διαλεκτικοῖσ το μεν 'επαγωγή 'εστι το δε συλλογισμόσ το δε φαινόμενεσ συλλογισμόσ, και...ὀμοίωσ 'έχει...."[23] Thus induction: " 'επαγωγή" is involved in dialectics, and it provides the basis for the apparent universality which these principles are supposed to reflect.

It is helpful to scrutinize Aristotle's conception of inductive inference at this point so as to appreciate the way it plays a special role in the evolution of dialectic. In his recent book, *Experience and the Growth of Understanding*, D.W. Hamlyn observes that for Aristotle induction meant that the "repetition of experiences of a similar kind leads to the setting up of a universal judgment." Thus, unlike a notion which explains induction as the formation of generalizations from the collection and examination of specific cases having something in common, the Aristotlean view emphasized the role of repetitive experience, which develops in the observer the "realization" of similarities and differences among the various particulars. Apart from the epistemological tenability of this thesis, which Hamlyn shows to be in need of clarification and improvement, Aristotle's position on induction presupposes the formation of patterns of recognition, which enables the investigator to make determinate the undifferentiated.[24] Hence, the generalization which results from this process of reasoning reflects the mind's capacity to abstract homogeneous characteristics from particulars. Within the context of setting forth the principles of dialectic, the inductive process of reasoning provides a means of discovering regularities in certain successful modes of disputation, and these uniformities are only as "intuitively certain" as their many confirming allow. Their universality then

---

[23] Aristotle, *Aristotelis Opera*, II, Academia Regia Borussica, p. 1356.

[24] Hamlyn, D.W., *Experience and the Growth of Understanding*, International Library of the Philosophy of Education Routledge and Kegan Paul, 1979, p. 61.

is based upon the fact that the mind is not faced with any contrary instances to whatever rule has been found instrumental in successful disputation.

The "rigor" one should anticipate here is one which results from observing successful procedures for sincere inquiry. It should not be expected that the same kind of rigor found in a fully defined system such as formal logic should also be evident in relationships of preference as well. As E. Weil points out, both syllogistic reasoning and dialectics deal with the form of reasoning. However, the difference between the two is that syllogistics is "the presentation of discovered truths in a logically water-tight way, "whereas dialectics" is a communal search for truth."[25] The validity of the latter (as procedure) lies in the fact that it has been found to work under tested conditions. Its rigor results from the fact that it has been observed during a sufficiently long period of time.

To insist upon precision similar to that of syllogistic reasoning here is to miss the whole purpose of developing a logic of preference. One can, of course, employ a formal language to describe the constituents of preferences, over spans of time, etc. Indeed, this is the valuable work R.M. Martin has performed in articulating the extensional features of expressions of preference. However, Martin claims that the formal descriptions he is offering should be taken as "first steps" at securing definitions of different facets of preference-relations between linguistic expressions, considered on a pragmatic level. Martin is keenly sensitive to the crucial fact that it is the formal language which is made to reflect the character of the object under investigation, and not conversely. Thus, one cannot expect a formalization of preference expressions to exhibit any more rigor than that which accrues to the observable phenomena which manifest them. In this respect, Aristotle's insight into the inductive aspect of the formalization of preferences stands as a solid contribution in this area of philosophical inquiry.

Having seen a parallelism between Aristotle and Martin in the area of a logic of preference, one may become susceptible to the oversight of not considering sufficiently the nature of the contribution Martin himself is offering in this domain. In this regard, one must naturally take into account the context in which his work appears, so as to assess its shift of direction and significance.

As noted above, Martin presents his formalization of preference within the framework of an experimental situation where some investigator is interested in seeing whether a subject finds a particular sentence more or less ac-

---

[25] Barnes, J., Schofield, M., Sorabji, R.,*Aristotle on Aristotle*,p. 107.

ceptable. Thus, the formalization of preference occurs as a result of observing and formalizing the function of language in a metalanguage, which assumes both the operations of deductive and inductive logic.

Martin goes to great lengths to explain the objective of his particular approach to the logic of preference in the following passage:[26]

> It should be noted again that the experimenter E never imposes any of these rankings upon X. He experiments and observes whether X's preferences fit the definitions or not. X is quite free to exhibit whatever preferences he chooses, and the rankings may or may not satisfy certain reasonable requirements of consistency or normalcy or rationality. If they do these various definitions will help E to classify X's preferences by placing them in the proper rankings. Under certain circumstances they might also be of help to X himself, in helping him to revise his preferences so as to accord with one or another of these ranking ..."

One sees here how Martin is very much aware of the fact that the formalization of preference can have a didactic effect, in that it can illustrate what one's ordering of preferences *should be*, given a sampling of preferences, so to speak. Of course, a great deal depends upon whether or not the respondent is willing to accept the logical consequence of the rankings he is shown. In this, it is important to consider how more so in Martin than any other writer considered thus far is the view expressed that there is no single definitive logic of preference which somehow one must try to discover and pin down. Rather, any formalization concerning preferences must be sensitive to the individual situation being scrutinized under experimental conditions. The best that can be done is to suggest very general rules governing the ranking of preferences, within a formal language L. Moreover, as he makes it quite clear in the above passage, the formalization of preference can only be descriptive, not prescriptive.

A striking insight which emerges at the outset of Martin's analysis of preference-ranking concerns his differentiation between "choice" and "preference." He characterizes the latter as a kind of extended historical event which is in essence a string of "continuous choices." Thus, from the outset, preference is seen by Martin as an event which is composed of a sequence of atomic choice events, which in turn constitute the broader molecular event representing preference.

---

[26] Martin, Richard M., *Intension and Decision*, p. 56.

In view of the above, he introduces 'Prfr' as a primitive in his metalanguage, which expresses a quadratic relation such that:[27]

'X Prfr a,b,t',

which says that X prefers sentence a to sentence b at time t. The incredible simplicity and clarity of Martin's approach cannot but be noted. Furthermore, it is a tribute to his technique that as he progresses to more complex preferential-relations the same economy of expression and clarity of insight is retained throughout.

In Martin's view 'Prfr' expresses a transitive and asymmetric relation between two possible sentences, a and b, so that if X Prfr a,b,t, then it is not the case that X Prfr b,a,t. This enables him to introduce the notion of *rational hegemonic ranking* quite simply as the case where given say three exclusively different sentences of L, the investigator can expect to observe any of precisely six preferential variations groupings with respect to these three sentences.

With the above, Martin proceeds to survey the prospect of introducing further types of rankings by injecting into the rational hegemonic ranking a relation of *indifference* which is defined in terms of equality of preferential status with respect to given sentences a, and b. Here one encounters the new ranking situation, where the investigator can expect precisely four possible preferential outcomes for the three possible choices a, b and c. For the respondent may prefer a to b, or he may be indifferent between b and c, or it may be the case that he is indifferent to c and b, or finally he may prefer a to c. Interestingly, the relation of indifference, which is seen to be both transitive and symmetric, does not really constitute the introduction of anything "new" in L, since it is seen to be defined in terms of the relation of preference. Here again, the pronounced economy of Martin's approach should be appreciated.

Allied with the notion of rational hegemonic ranking, Martin sees the option of having what he terms rational preference rankings. The latter is simply a way of ranking sentences, say a and b, given the assumption that either a is preferred to b, or b is preferred to a, or that the respondent is indifferent to both a and b.

Within L, Martin notes the possibility of having rankings which reflect the use of logical constants, such as *vee* and *tilde*. Assuming as well the operation of an inductive logic in one's metalanguage, the way is open for rankings which reflect an ideal respondent assessing "the degree of confirmation on suitable evidence" for a or b. Thus, such preference-rankings would never be in error given that the "correct" probabilities have been considered.

---

[27] ibid., p. 47.

His "General Preference Rules" were discussed in connection with the comparison with Aristotle, and much more need not be brought out at this point. Nonetheless, it is worth mentioning that Martin is articulating his formalization of preference on the basis of a theory of time which is entirely extensional, and involves a notion of identity. Similarly, his formalization of "rational preference-ranking" was also discussed above in connection with Aristotle. What is important, however, relative to the notion of a rational preference ranking, is Martin's introduction of symbols such as 'F' and 'G' in the definiendum, standing as "expressional abstracts of the metametalanguage."[28] This is to say that his definition of rational preference-ranking involves realizing that the investigator X is dealing with rational preference-ranking of some virtual class of sentences, expressed by 'F'. Thus, through the use of logical techniques, Martin is able to achieve a depth of expression ranging from an object language to a metalanguage, to a metametalanguage.

The flexibility of Martin's technique is again illustrated where he extends his investigation to include some additional preference-rankings. For example, where he sets forth the *normal preference-ranking* for virtual class F, he specifies that preference-ranking in this case will be what is normal with respect to the use of the *tilde*, as applied to "whole" sentences. Here one assumes the acceptance by the respondent of the syntactical rule governing the *tilde*. It should be observed how easily Martin introduces into the formal analysis of preference this aspect of the syntactic, which now constitutes a primitive element of L. Interestingly, it is the "use of the tilde" which is being supposed in L, and thus it is introduced by the italicized *"tilde"*. Ranking in this case is presented in standard Martinian style as follows, assuming that one is dealing with the *use* of *tilde* for two different sentence:[29] 'F NPR *tilde*, X, t' abbreviates $((Ea)(Eb)(Fa \cdot Fb. \sim a=b. F(\textit{tilde } a) \cdot F(\textit{tilde } b)) \cdot (a)(b)((Fa \cdot Fb. \sim a=b. F(\textit{tilde } b)) \supset (X \text{ Prfr } a,b,t \equiv X\text{Prfr}(\textit{tilde } b) . (\textit{tilde } a),t)))'$. The preference-ranking in this case is such that where a is preferred to b at time t, then the negation of b is preferred to the negation of a, at time t.

The applicability of such rankings is seen by Martin with respect to the ideal situation noted earlier, were it is supposed that some respondent X is capable of knowing correctly the degree of confirmation c, "on suitable evidence", e, for some sentence a. For example, suppose X's preference of a over b is based upon the fact that the degree of confirmation for a over b is such that:

---
[28] ibid., p. 49.
[29] ibid., p. 54.

$c(a,e) > c(b,e)$. Consequently, based upon the preference-ranking relation for *tilde* defined above, and the evidence for *tilde b* and *tilde a* standing as follows: $c((tilde\,b), e) > c((tilde\,a,e))$, then it can be said of X that he prefers (*tilde b*) to (*tilde a*) at *t*, if and only if he prefers *a* to *b* at *t*.[30]

Though the preference-relation for *tilde* is presented as a general ranking rule, Martin is quick to point out its immediate applicability in the case involving assessing the degree of confirmation for two distinctly different sentences *a* and *b*. His introduction of the syntactical rule for *tilde* gives all that much depth to his formalization while not compromising the openness of his analysis as far as its capacity is concerned in reflecting the many different modes of preference-ranking. In this case of *tilde* ranking, Martin is willing to admit that there may very well be an instance where the respondent simply is not aware of the rule for *tilde* and may not recognize the implications of his preferring *a* to *b*. Still, the point can be made that X may be ranking his preferences along the lines of the rule for the *tilde*, though he may not be aware of it. In such a situation, Martin points out that the above rule for ranking preferences is no less valid, it is just that some other term than *tilde* must be used to express the kind of relation attributed in logic by the *tilde*.

Martin proceeds to offer a second preference-ranking rule, where this time what is being assumed is the normal logically correct use of *vee*, for disjunction. Now the abstract class of sentences, expressed by 'F', contains sentences *a*, *b* and *c*, so that where X prefers *c* to (*a vee b*), at *t*, then X prefers *c* to *a* and *c* to *b*, at time *t*. The normal-preference ranking rule with *vee* relative to the abstract class of sentences F is presented very straightforwardly as follows:[31]
"'FNPR *vee* X,*t*' abbreviates $((Ea)(Eb(Ec)(Fa \cdot Fb \cdot Fc \cdot -a = c \cdot (-b = c \cdot -(a\,vee\,b) = c \cdot F(a\,vee\,b))\,(a)(b)(c)((Fa \cdot Fb \cdot Fc \cdot -a = c \cdot -b = c\,.\,-(a\,vee\,b) = c \cdot F(a\,vee\,b)) \supset (((X\,Prfr\,a,c,t \,v\, XPrfr\,b,c,t) \supset X\,Prfr\,(a\,vee\,b),c,t \cdot (X\,Prfr\,c, (a\,vee\,b), t \supset (X\,Prfr\,c,a,t\,X\,Prfr\,c,b,t))))$'"

The definition of normal preference-ranking with respect to *vee* simply says that given the difference of sentences *a* and *b* from *c*, and that X prefers *a* to *c* at *t*, or that X prefers *b* to *c* at *t*, then X prefers (*a vee b*) to *c* at *t*, and if X prefers *c* to (*a vee b*), then X prefers *c* over *a* at *t* and X prefers *c* to *b* at *t*.

Again, Martin notes how even with this definition one can introduce considerations dealing with the degree of confirmation for sentences *a*, *b* and *c*, relative to suitably available evidence *e*. Without introducing a needless repeti-

---
[30] ibid., p. 55.
[31] ibid., p. 55.

tion of material, Martin's point can simply be presented as where if the respondent is seen to prefer a to c at t or b to c at t, then the degree of confirmation, c, for a or b relative to c is such that $c(a,e) > c(c,e)$ or $c(b,e)$ b)$,e) > c(c,e)$. On the other hand, where X prefers c to $(a\ vee\ b)$, then the degree of confirming evidence e for c relative to a and b must be such that $c(c,e) > c((a\ vee\ b))$.[32]

Martin proceeds to suggest in a summary fashion a few additional ways in which preference-ranking can be formalized, as for example where X is found to prefer logical theorems as opposed to false statements. Consequently, one could be said to have a preference-ranking with respect to logical theorems of abstract class F. Moreover, one could have a formalization of preference-ranking relative to X preferring theorems as opposed to falsehoods. In the latter case, one would not have X preferring theorems to nontheorems, for it may be the case that F might contain a sentence which is *undecidable*.[33]

The sense of "normal" is recognized by Martin to vary in the above definitions of ranking, since in some cases it is said to reflect the degree of confirmation on evidence for a sentence, whereas in other cases it is said to reflect the use of *tilde* or of *vee*. In all of this, Martin cautions that these definitions are not intended to be complete, but rather are suggested as tentative examples of how the formalization of preferential expression can be had, within the context of experimental observation.

Given the historical sequence in which Martin's work emerges, one cannot help but note the profound depth of his contribution. His sensitivity to the syntactic, semantic and pragmatic dimension of the preferential is simply not to be had with the same degree of clarity in the work of any other analyst up to his time. The tedious intentional mode of analysis encountered in von Wright's *The Logic of Preference* is circumvented in Martin's extensional approach. This is to say that von Wright's often highly involved reference to possible world states, and how these influence the development of a logic of preference, is noticeably absent in Martin's approach. For this reason alone, it seems quite evident that the latter provides a more streamline approach to the issue of developing such a logic, though without sacrificing depth and power in the process.

A point of general comparison is in order between von Wright and Martin since both of their works appeared in 1963, and their respective approaches

---

[32] ibid., p. 56.
[33] ibid., p. 58.

to the analysis of preference are so widely different. For instance, Martin treats preferences in terms of their extension, and therefore *extrinsically*; von Wright, it will be observed considers preferences in terms of their intention and therefore *intrinsically*. The latter was also seen to distinguish between preference and choice, particularly where choices involved the probability of risk. Indeed, in the sense adopted by von Wright, preference is regarded quite independently of any type of choice, since a "personal liking" such as preferring sunshine to rain is separate from anyone's choice in the matter of these two alternatives. By contrast, Martin's approach is less in fear of the novelty and/or variety which spontaneous choice may precipitate in the determination of preference. As already noted, it is the observation of a history of choice selections which forms the observer's judgment that such and such expresses the respondent's preference. Characteristically, Martin's endeavor is designed to be descriptive of the preferring state. On the other hand, von Wright presents a general value-theory which is intended to prescribe the correct order of preferences, considered as given. Thus, his *ceteris paribus* clause serves to insulate his theory from any and all empirical concerns.

Moreover, it is interesting to compare von Wright's conception of the state of affairs which the preference-relation involves, with Martin's manner of conceiving of the constituent elements of this relation. Von Wright sees 'p' and 'q' as expressing "proposition-like" entities, or more specifically they are used to symbolize sentences describing generic states. Though he is quite clear in his insistence that the states which preferences deal with must involve persons, occasions and segments of time, von Wright never goes beyond the specification of 'p', 'q', etc. as symbols for sentences. Martin, however, because of his adoption of an extensional semantics, shows with markedly superior precision the constituents of the sentence expressing the preference, in terms of object, property, time span, etc. as these relate to the language user expressing or uttering the sentence dealing with the preferred state.

However, apart from the clarity of Martin's contribution to this field, a number of serious questions emerge as to the degree in which his formalization reflects a "general" view of preference. It is apparent that Martin concentrates upon a conception of preference which is manifested largely within a scientific and pragmatic context. In this, his view is divorced from any context involving the moral suasion of the "intrinsically good." Consequently, von Wright's investigation perhaps has a more difficult task to accomplish. Yet, it can be asked whether Martin's approach to the subject is perhaps not too confined to a very limited area where choices, and the preferences which emerge from choices, are governed by the scientific method of experimentation. In a sense, it must be realized that however tentative Martin's remarks may be, and however much he advocates an inherent plasticity in any attempt to formalize preferences, so that there can never really be a *single* logic of preference, still the framework of experimenter and respondent which dominates his analysis, must operate in some way to circumscribe the logic of preference he is suggesting. This seems to be evident in that aspect of his work where he observes that perhaps the respondent is not aware of the fact that his preferences can be characterized in terms of the *tilde*, nonetheless perhaps another term can be used which is comprehensible to the respondent in place of the logical terminology used to express the *tilde*. In this, there is a trace of the belief that even in "ordinary" circumstances where the individual is not aware of some logical operation, still his preference "may" be said to conform to the structure which these notions of logic provide. However, one is reminded here of the insight provided Kenneth Arrow, to the effect that it is often the case that even though a is preferred to b, and b to c, it need not "logically" follow that a is preferred to c, in the ordinary affairs of human beings. Thus, at most, Martin's formalization can provide a means of formalizing the language of preference within the restricted field of scientific inquiry. Much remains unsaid, however, concerning the applicability of this approach to preferences in everyday circumstance, where "logical" consequences need not be of the utmost importance.

The comparison between Martin and von Wright receives an added significance where one considers more closely the nature of the difference between the former's analytical approach as opposed to the latter's. In Martin's case, the extensional pragmatics of preference is evolved from a mode of inquiry which brings forward the fundamentally ostensive character of the objects under investigation, in this case the *denotata* comprising the preferring of some object or state of affairs, as a sentential expression. In von Wright's case, the emphasis is also upon what is preferred though *as a sentence*, loosely conceived as an "entity-like" proposition, with no concern as to the denotation of the expression involving what is preferred. As can be seen from a careful examination of von Wright's contribution, one of the major difficulties with his logic of preference is that it becomes a logic where the modality of preference is handled as a modifier of sentences or propositions about what is preferred, instead of it being a modality about *things* which are preferred. Thus, von Wright's logic assumes a modality which is *de dicto*, instead of it taking a more plausible view of preference, which would be one of *de re*. In a sense, it can be said that Martin's mode of inquiry transcends the sentential surface, so as to fathom the ultimate denotative constituents of the preferring act, conceived ultimately as a spatiotemporal event. Von Wright, on the contrary, remains on the surface of the expressed preference itself, and treats the content of the preference as an impenetrable level of meaning, curiously alluded to as at times a sentence or a proposition, or as a "proposition-like" entity. Though one can readily identify the obscurity in von Wright's manner of conceiving of the object of his investigation, this is not to say that Martin's endeavor is totally void of difficulties. Somehow, where the expression is dissected into its basic denotative components, one wonders how much of the analysis pertains to the "ordinary" sense of preference. The increased referent specificity seems to be at odds with justifying the outcome as what is meant in the *everyday* meaning of preference.

Relative to the above, there is a further question with respect to what Martin is proposing where he says that the formalization of preferences he is suggesting can be applied to objects, and not just to "whole" sentences. Toward the close of his presentation on the formalization of preference, he claims to see no problem in applying his conception of the preference-relation "Prfr", as well as all his definitions dealing with normal preference-ranking, to objects or "nonlinguistic" entities. As an example of this, he claims that one could say that respondent X prefers a to b at time t, where say 'a' stands for a cup of coffee, and 'b' a cup of tea, and where it is understood that the preference expressed is *"with respect to such and such a property P."*[34] The property alluded to here is the having of a certain flavor, so that the expression

$$X \text{ Prfr } 'Pa', 'Pb', t$$

says that X estimates that a has a certain flavor P as opposed to b having that flavor P at t. More simply, the above is saying that X prefers a to b at t with respect to the property P.

In the context of this study in general, it was seen that if it is going to be maintained that Aristotle's pronouncements concerning preferences dealt solely with things, then Martin's point above serves to draw closer the parallelism between his work and that of Aristotle's. As it turned out, however, it was seen that aspects of Aristotle's analysis can be seen as involving preferences as expressed in some sentential form, and thus a comparison is possible with Martin more along the lines of Martin's initial intention of devising a logic of preferences, dealing with sententially expressed entities. However, Martin's laconic remarks at the end of his presentation are not fully convincing with

---

[34] ibid., pp. 59-60

respect to the fact of whether he can make the transition from relations involving preferences as sententially expressed, to the formal expression of preferences involving nonlinguistic objects. In the example presented above, for instance, how congruous is it to claim that one is offering a conception of the preferential-relations which is presumedly extensionalized and pragmatic, and which is now said to be operating in expressing X's preferring the *flavor P* of a to that of b? Does it make sense to say that the flavor is somehow part of the object, be it coffee or tea? How does 'P' designate flavor? It seems that Martin is suggesting that somehow the property P is separable from the object, be it coffee or tea. Yet, however one wishes to construe this aspect of the above, it does little to explain how one is to handle extensionally the "having of the flavor" expressed by P. For in the way Martin presents his example, 'P' simply does not designate a property, but rather the *having* of the property. In essence, he says that X estimates that a will exhibit property P, as opposed to b doing so. Still, apart from the introduction of X's estimation of what a and b are like relative to P, in the case of a flavor it is the person doing the savoring who is said to *have* the flavor, and it is not clear how the property can be extensionalized as anything other than a sentential report that X has, or anticipates having, property P relative to object a, as distinguished from his having this property relative to object b. Without belaboring the point on what may have been an ill-chosen example, Martin's commitment to the scientific approach of testing to determine preference entails the examination of reports by a respondent which may be a way of locking his formalization of preference into a context which requires allusions to linguistic entities alone. Consequently, the transition he is supporting to cover preferences which deal solely with things as nonlinguistic entities appears to conflict with certain fundamental presuppositions of his method of investigation at the outset. His introducing the qualification that a is preferred to b "with respect to property P" simply disguises an embedded sentential entity residing in the background of his analysis at this juncture. For with the introduction of the qualifying phrase he seems to be alluding to a as an *a thing under a description*, which implies once again a sentential expression. If this implication were not present, then it is difficult to comprehend the function of his relational-expression 'Prfr'. Finally, in this connection it should be recalled that Martin's commitment to the technique of scientific investigation and testing requires in most cases a framework of some theory or hypothesis in which propositions, as sententially expressed entities, are tested and evaluated. Thus, a formalization of preference, primarily articulated in such a context, can hardly be conceived as having dealings with objects alone, in complete isolation from an exegesis. All in all, von Wright's perceptive though laconic observation that an analysis of preference involving things must be fundamentally an analysis of preference concerning sentences about states of affairs may be in the end quite sound.[35]

However, one must keep sight of the fact that Martin presents his suggestions concerning the suitability of applying the ideas he presents concerning the preference-ranking of sentences to the preference-ranking of objects in a highly tentative fashion. In the absence of further elaboration on his part, it is unfair to argue that the transition he is advocating is totally without foundation. What is needed, no doubt, is further analysis as to how the extensional and pragmatic approach can be shown to apply to things, and not only to sentences about things. Martin's endeavor to secure the expansion of his perception of preference-rankings is generated by his belief that there may be very productive results if this can be shown in the case of economics. Though this may still prove to be true, very much remains by way of exegesis to illustrate exactly how this is to be done.

Looked at in the context of the other efforts in this field to effect a viable formalization of preference-relations, Martin's work stands out as an important achievement. Its closeness to the "practical" state of affairs wherein a preference is encountered is not emphasized in any earlier attempt. Furthermore, it proceeds along the lines of not preconceiving what preferences are or should be. Rather, the ordering of preferences results from testing and observation, whereby the results observed are considered in terms of a formal structuring reflecting semantico-pragmatic considerations. This is in sharp contrast to Ramsey's approach, for example, which begins by presupposing that persons act in a way which will always secure the object of their desire. Consequently, whatever Ramsey has to say about preferences is guided by this fundamental presupposition. In essence, Martin's attempt tries to leave open the novelty and variety of individual choice, while illustrating the possibility of subjecting to formalization any consistent pattern of choice which that person may manifest in his observed behavior.

Martin's major critics on this heading, notably Richard Montague and John M. Vickers, question his consistency when referring to the cognitive and evaluative aspect of preference. Montague, for example, charges that Martin fails to distinguish between value and belief, arguing that there are times

---

[35] Von Wright, Georg H., "The Logic of Preference Reconsidered," *Theory and Decision*, vol. 3, (1972), p. 144.

where Martin's "Prefers" is taken to mean either 'believes more strongly', or 'values more highly'. Moreover, he accuses Martin of inconsistency where the former alludes to 'degree of acceptance' as meaning sometimes 'degree of belief' and at other times 'subjective utility'. On Montague's interpretation, Martin is wrong in thinking that utility and belief have the same formal properties. Along these lines, Montague pursues Martin where the latter allegedly argues that one's preferring Titian to Orozco as a painter of nudes is equivalent to saying that one *believes more strongly* that Titian is a fine painter of nudes than that Orozco is a fine painter of nudes. Montague claims that someone may believe with *an equal degree of intensity* that Titian is a fine painter of nudes, and that Orozco is a fine painter of nudes, while preferring Titian to Orozco.[36]

Vickers adumbrates Montague's reservations concerning Martin's work, and proceeds to push matters even further. In the quote cited above from *Intension and Decision*, (footnote 8), Vickers finds the most damaging evidence, claiming that it fully exhibits the supposed confusion between preference and belief, recognized earlier by Montague. In Vickers' view, Martin's unpardonable sin is that of his being totally insensitive to the uniqueness of the preference-relation as one which has its own identity, so to speak, and which thus should not be conflated with other concepts, e.g., belief or acceptance. In underscoring this criticism, Vickers cites the von Neuman and Morgenstern attempt to explain preference-relations in terms of expected-utility. In such attempts, efforts are made to separate objective probabilities expressing expected utility from subjective probabilities, which are an agent's personal belief in the likelihood of an event occurring. Where these two factors are not distinguished in this and other theories, it becomes virtually impossible to achieve any clear formalization of preference-expressions.[37]

Both critics seem unusually contented with a surface reading of Martin's text, considering nothing of the highly restricted context in which he is operating. At first glance it seems unproblematic to counter Montague's point on the Titian-Orozco example. Martin would agree that the same agent could believe both that Titian and Orozco are fine painters, this is not a contestable point. However, what if that same agent is asked to choose between these two painters. Martin's solution would be that preferring Titian to Orozco is tantamount to the agent's believing to a greater degree that the former is a finer

---
[36] Montague, Richard., "Intension and Decision," (review), *Journal of Symbolic Logic*, Volume 31, 1966, p. 101.
[37] Vickers, John M., "Intension and Decision" (review). *Journal of Philosophy*, Volume 64, 1967, pp. 197-198.

painter than the latter. This translates in terms of the extensional and pragmatic characterization of the agent's willingness to pay more for one painter's work as opposed to the other's, or to defend one painter's work at the expense of the other's, or perhaps even to display the former's work more prominently than the other's, etc., etc. The essential point here is that Martin is operating wholly within a context of testing and experimentation. This is a highly specialized frame of reference; and from the Peircean tradition from which he is launching his investigation, it is a perfectly consistent and unambiguous interpretation of the concept of preference. In the final analysis, Martin is saying that preference, within this context of testing, means *what* one is willing to do to insure the realization of some state of affairs or event, sententially expressed as a true assertion concerning things in the physical world. Montague's charge of Martin's confusing preference with belief would hold only if the latter's avowed aim was to present a general account of belief and preference in every conceivable sense. This is far removed from Martin's stated purpose.

Chapter 3.

*Rescher's Logic of Preference and Linguistic Analysis**

Professor Rescher ranks along with Henrick von Wright and S. Hallden as one of the pioneers in the attempt to evolve a logicized semantics of preference. Though influenced by von Wrights's ideas concerning states of affairs and how these are useful in explaining certain formal implications between preferences, it is Rescher who introduces the method of attaching numerical units of merit to possible world state descriptions reflecting preferences. His aim in doing this is to arrive at purely formal distinctions between various expressions of preferring. Thus, preference-principles are found by Rescher to have the same rigor as that of arithmetical truths, since they are considered apart from any particular or *synoptic* contexts.

Rescher's overall purpose in proceeding as he does is to bridge the gap between the "logico-philosophical" tradition for a logic of preference, which was manifested in Europe from the early 1900's, and the predictive "mathematico-economic" tradition which underlies recent decision and game theories. His analysis leads him to an early rejection of a purely intuitive and axiomatic method for devising such a logic, as is found in the works of von Wright, Chisholm, Sosa, and R.M. Martin. Alternatively, Rescher pursues a strictly semantic approach, with a view toward illustrating the mathematical necessity which he believes the intension of preferential discourse translates into, once a context of possible preferential state descriptions is posited, as is done in economics.

The aim here is to investigate the soundness of Rescher's approach, insofar as it employs a mathematical model to elucidate a linguistic phenomenon. First, certain incongruities are noted between Rescher's purely linguistic conceptualization of preference expressions and Alan R. White's re-

---

*Segments of this chapter were first published in *Logique et Analyse,* and are reproduced here with the kind permission of the publisher. The present chapter constitutes a refinement of issues and arguments presented earlier.

vealing remarks concerning the grammar of discourse involving the modality of needs, which seems to underlie preferences. Secondly, Rescher's method of attaching numerical merit to states of affairs is not found to reflect the kind of "mathematical rigor" among expressions of preference, which he believes he has demonstrated. These observations lead to a third and broader discussion concerning the usefulness of formalization, and whether talk about preference defies quantitative analysis, at least in the way suggested by Rescher.

Though problems can be recognized in Rescher's presentation, it must be said that the value of his work as an exploratory effort is without question considerable.

I.

"... The study of preference-principles acceptable upon abstract, formal, systematic grounds rather than upon any particular substantive theory of preferability-determinations is the task which the philosophically orientated "logic of preference," as we envisage it, is to set for itself."[1]

In the above quote Rescher tries to emphasize his interest in the formal characteristics which expressions of preference exhibit. To capture the purely formal character of preference, he predicates his remarks on determining in some suitable way "a numerical measure of merit: $\mu(a)$", so that the evaluations of preference, expressed propositionally as: $\mu(a)$, would be identified and ordered according to their priority. Rescher observes that his particular approach of using numerical merit can be substituted by another modality, such as that of the desirable, undesirable, etc. and the results would be the same.[2]

His remarks concerning how modalities such as the desirable, etc. can be substituted univocally in place of the method of attaching numerical merits reveal a basic assumption relative to his interpretation of preferences. Namely, he sees preferences as some economist would, i.e., as having value, or of being valued, because of some intrinsic quality they presumedly possess. For this reason he finds no difficulty in introducing the modality of desirability as an alternative means of expressing the evaluative aspect of preference. The desirous is considered by him as intrinsically desirous, much in the way Moore

---

[1] Rescher Nicholas, *The Logic of Decision and Action*, University of Pittsburgh Press, 1968, p. 38.
[2] ibid. pp. 45-46.

speaks of the intrinsic goodness of things. Hence, he sees the attaching of numerical merit to a preference, or the determination of whether it is desirous or not, as alternative ways of expressing the value of the preference which somehow inheres as given within the preference itself.

The assumption of taking preference as having some inherently intrinsic value permeates Rescher's important distinction between *first order* preference and *differential* preference. In the former case, preference is considered simply in terms of the *immediate* effect upon someone when a particular state of affairs comes about. For example, where the preference is the gaining of $1, then one is better off, should this event occur, because the effect of the event is something the individual would want or should seek out. In Rescher's view there is no question as to the intrinsic goodness of gaining $1. Similarly, the notion of intrinsic value plays a role where Rescher speaks of differential preference. In the case of the latter, the preference for the lesser of two evils, say, losing one dollar to losing one hundred dollars, is still a preference for something which "under the circumstances, the occurrence of losing of one dollar is *itself* (my italics) a very good thing ..."[3]

Rescher is aware of the traditional problems which the notion of "intrinsic" goodness or value entails. Yet he feels constrained to use it as a means of pursuing his "semantic" analysis. Significantly, his use of the term "intrinsic" is consistent with his adaptation of the methodology of economists, who employ it so as to arrive at an ordinal indexing of preferences conceived as expressions of an individual's desire for or his wanting of some commodity. Indeed, it is their use of numerical ordering which inspires his own attempt to introduce numerical units of merit as a means of differentiating between various expressions of preference. However, it is his adaptation of this idea which will be at issue here, as well as the allied presupposition that preference is somehow related to the wanting or desiring of an object.

In broad outline, Rescher's theory of preference assumes that preference-relations manifest the same rigor mathematical relations do. This is to say that preference, whether taken in the *first-order* sense of someone preferring a to b, or preference in the differential sense of preferring a to b, under say condition c, exhibits relations having the kind of rigor encountered in mathematics. This is not to claim that first-order preference relations can be correlated to differential preferences, but rather that for Rescher the two can be seen to have a clearly specifiable precision. Rescher's semantic theory thus involves

---

[3] ibid. p. 40.

the essential notion of a "*propositional preference ordering*," sufficiently powerful to interpret both first-order and differential preference. The mathematical rigor he insists upon is derived from first assuming that certain possible world states have an index of *merit* measure. In essence, here Rescher is simply showing how the ultimate logic of preference he desires to derive is parallel to the notions involving different types of *goodness*. With merit assumed as ascribed to possible world states, Rescher proposes the attribution of numerical units of measure to truth-functional propositions, which are taken as compounds of possible world states. The number ascribed to these propositions is itself an average of the possible worlds w_1, for which the said proposition is said to hold true. Rescher by-passes any discussion of the material reasons for the attribution of this merit value, concentrating solely on the formal consequences of such attribution relative to the value of propositions. Within this framework Rescher goes on to present "preference-tautologies," which are then shown to be reducible to arithmetical truths.

Preference is not generally conceived of as an object in the physical sense, but as a *relation* between an individual and some particular state of affairs. One sees, however, that Rescher employs a methodology found in economics to develop his version of the logic of preference which emphasizes *only* the objects which preferences involve. Economists attach values (i.e. prices) *to* objects in the context of supply and demand as determined within a market place exchange. However, where one is concerned with the analysis of preference as a linguistic phenomenon, the issue of whether one is dealing *simply* with physical objects is highly debatable. For it can be seen that, though desires and wants are said to play a role in the determination of pricing, economists merely mention them without going on to explain how they play a role in determining preferences. Hence, to what is Rescher attaching the units of merit: to the objects of desire or want, or to one's wants and desires themselves in *relation* to these objects? The former alternative would lead him toward adopting a strictly extensional sense of preference, which is something he explicitly avoids in the body of his presentation. The latter alternative, involving attaching numerical units of merit to subjective states in relation to objects, raises the foreboding issues of establishing criteria in a private language, spoken of by Wittgenstein in the *Philosophical Investigation*. Consequently, these immediate difficulties beset Rescher's fundamental procedural assumptions regarding whether one can attach numerical designations to objects of desire and want in the manner analogous to that found in economics, while also alluding to the intrinsic desirability of the values of that which is preferred.

Interestingly, Rescher often relates the idea of preferring to desiring or wanting, much in the way of suggesting that the logic common to the latter can be used to elucidate the formal relations of preferences. In light of Alan R. White's insightful remarks in *Modal Thinking*, however, it is important to see that one does not speak of wanting or desiring to prefer. In an important passage, White says it makes no sense to want: "... to expect, *prefer* (my italics), fancy, imagine, regret, envy, dread or mind anything, but it makes sense to say that one needs (my italics) or does not *need* (my italics) to do any of these ..."[4]

To understand the direction of White's thinking and how it pertains to Rescher's position, one must first introduce the basis of White's view as it relates to the distinction between 'lack' and 'want'. White sees a confusion in supposing that one *always* wants that which he does not have. This confusion is based upon believing that 'wanting' is the same as 'lacking', which is thinking that "to want is to want to get". He points out that one can want that which he has, as for example, one can want his car so that he will not have to lend it to someone else.[5] Hence, his wanting that which he has does not introduce the idea of 'lacking' into the picture.

Rescher is unconcerned with the difference between 'lacking' and 'wanting', at least with respect to the development of a logic preference. For he sees his approach to the formalization of preference as "... an *evaluative* one, in which preference-relations are based derivatively upon an essentially *quantitative* approach, the assessment (measure) of the intrinsic merit (goodness) of the objects involved".[6] In this passage, the underlying assumption is clear, namely, valuation necessary for determining the formal relations of preference is derived from "possible world" situations which are considered to be desired and meritorious according to their "intrinsic" value. His thinking here is apparently somewhat similar to G.E. Moore's perception of intrinsic value, in a theoretical sense.[7] For the latter, intrinsic value can be abstractly explained in terms of securing or attaining invaluable pleasure. In essence, Moore argues that hypothetically states which contain an *excess* of pleasure over pain are seen by some as intrinsically valuable.[8] Consequently, given how most hu-

---

[4] White, Alan R., *Modal Thinking*, Cornell University Press, Ithaca, New York, 1975, p. 113.
[5] ibid. pp. 114-116, and p. 109.
[6] Rescher, Nicholas *The Logic of Decision and Action*, University of Pittsburgh Press, 1968, p. 46.
[7] Moore, Georg, E. *Ethics*, Oxford University Press, New York, Chapter I.
[8] ibid. pp. 8-9, and pp. 100-102.

mans seek to avoid pain and attain pleasure, and the attainment of pleasure is usually the result of difficult effort which yields momentary gratification, one finds the whole concept of intrinsic value *in general* (and thus that of the evaluation of preference relations for Rescher) entwined within the notions of 'want' and 'lack'.

Significantly, Alan R. White goes on to observe that there is a difference between 'want' and 'need', primarily in the way that 'need' always seems to involve a context of instrumental end. White points out how the grammar common to the word 'want' is radically different from that which is common to 'need'. For example, expressions of 'want' do not necessarily carry along with them a reference to an end-state, whereas talk about needs invariably does involve such a reference. Expressions involving needs come to suggest some sort of constraint standing in the way of an end, whereas this is not the case where one speaks of wants. Again, only animate creatures can be said to want, whereas anything may be said to need; e.g. "morale" may be said to need a boost, as well as some professor's salary. Finally, one can alluded the fact that since all cases of wanting are confined to animate objects, it is involved with one's personal beliefs and perceptions of the world. The same is not the case with respect to needs. The latter requires demonstrable considerations dealing with "real" things that are "in the way of" something happening.[9]

Yet in arguing for the adoption of some of White's insights when trying to clarify the meaning of preference, one must be selective in choosing which aspects of his analysis of need are relevant and which are not. For preference is understood here solely as an expression of human concern toward the manifestation of a particular state of affairs. Thus, it would be wrong to argue that inanimate things prefer, for example, just as it would be difficult to justify the view that animals prefer. Hence, where White speaks of sub-human or inanimate things needing, these would be aspects of the meaning of need which does not pertain to the meaning of preference. Still, one would want to preserve those aspects of the meaning of need which seem to be very basic to the understanding of preference, such as where White sees needs in terms of their expressing a constraint standing in the way of a goal or an end-state, or as involving some sort of instrumentality.

It is interesting to observe that Rescher begins his paper by quoting from Aristotle's *Topics*, Book III. According to Rescher, Aristotle was the first to consider the problem of how preferences are to be analyzed. Rescher mentions

---

[9] White, Alan R., *Modal Thinking*, p. 103.

casually the term αἱρετώτερον used by Aristotle to speak of preferences. However, the αἱρετώτερον has a remarkable resemblance to all that Alan White says concerning the notion of 'need'. For Aristotle speaks of preference in terms of securing or seizing that which stands in one's way to an object. For example, Aristotle considers rational thought to be a state more preferable than that of pleasure or wealth, because of the greater permanence of the reflective faculty as opposed to the latter two states of being. Aristotle justifies the rationale of his observation on account that a disciplined mind will be instrumental to securing both wealth and pleasures, whereas the reverse is not a surety for the development of the intellect. One can observe this same reference to the instrumentality of preference αἱρετώτερον in Plato's *Philebus* where Socrates says that it is "better" (more preferable) to have understanding of the values of various tones and how they can be put together *in order to* create music. Thus, it is observed also in Plato how the preferable case is the one where something has been secured or grasped so as to be of *use* in attaining some higher end.[10]

The role of one's *needing* to prefer is quite strong in both excerpts, and it reflects White's idea that preference must have some connection with objective conditions and determinable states of affairs which must be mastered or altered in some way. The idea of instrumentality is thus basic to that which is preferred for these classical writers, and inner states of desire really have little to do with the notion of preference as found in their works. Rather, they are suggesting that somehow the preferable must be connected with the rational, wherein the latter involves an analysis of conditions and obstacles in securing a goal.

Evidently, the modalities of 'want' and 'lack' do not take into account a broad spectrum of consequences for states of affairs, whereas expressions of 'need' do. Furthermore, if 'need' is the modality which creates the context for expressions of preference, the latter should be looked at cumulatively, i.e. in terms of the actions which some particular agent must perform relative to a given state of affairs. Instrumentality thus seems to be basic to an understanding of preferences and their logic. To see discourse involving preferences in terms of wanting and desiring is to miss the actual complexity of what is being investigated.

---

[10] Aristotles, *Opera Omnia*, Vol. 1. 1973. Georg Olms Verlag, Hildesheim, New York, ΤΟΠΙΚΟΝ Book III, Ch. I, p. 197. Plato, *The Collected Dialogues of Plato*, Edith Hamilton and Huntington Cairns, eds., Princeton University Press, 1969, Philebus 11. 16c - 17d

Whether he is considering first-order preferences or differential preferences, Rescher does not allude to the role of "needs". He considers preferences solely in terms of how one is affected by certain states of affairs. This tends to make preferences into occurrences which *happen* to people, rather than situations which reflect an individual's active part in an environment, irrespective of the role of desires and wants.

The implication of White's position on Rescher's view is quite significant, since it raises the question of whether the logic of discourse which deals with pursuing the intrinsically good can ever serve legitimately as a model for a logic of preference. Robert Ackermann roughly perceives the problem with Rescher's position in that he sees how Rescher does not consider the role of *purpose* in preference. For example, one can prefer a Cadillac to a Volkswagen on account of the comfort which the former offers, or the reverse may be the case if one seeks economy of operation. Hence, the asymmetry and intransitivities which Rescher accepts as implicit in his notion of first-order and differential-preference do not seem to reflect sufficient flexibility to capture the complexity of the notion of preference.[11]

Ackermann's criticism does not touch the precise problem which limits Rescher's thesis, in that he does not see that it is the individual's needs which create the context for understanding preferences. For Ackermann, the idea of purpose is taken in a very broad sense, and he goes on to despair at the eventual evolution of a logic of preference because of the as yet unsatisfactory analysis of the idea of human purpose.[12] Surely, White's insights contribute toward appreciating the depth which the notion of preference presupposes.

Furthermore, the concept which seems crucial to the development of the logicized semantics of preference, namely 'need', complicates the issue of how one is to attach numerical value to state descriptions. Surely, if 'need' is the basis of preference, then one must consider whether 'need' can be measured at all. At least, this question is to be answered first, if one is to proceed according to Rescher's method of determining the numerical value which underlies preference relationships.

The problems which the above question involves turn out to be formidable. For one can feel a *pressing* need to eat rather than to listen to Chopin. Does one determine the *importance* of the need in terms of *what* is needed, or is one to

---

[11] Ackermann, Robert, "Comments on N. Rescher's "Semantic Foundations for the Logic of Preference", in *The Logic of Decision and Action*, Nicholas Rescher, ed. pp. 71-72.

[12] ibid. p. 71.

count the *intensity* of the need as a purely psychological occurrence? The latter alternative raises serious difficulties regarding the establishment of criteria for a private language. The former raises equally difficult issues. For example, which aspect of the need, if any, will be selected and assessed as possessing the merit? Moreover, it can be argued along with White that a need by itself is a state of affairs which does not allow for degrees, meaning that once something is recognized as a need, then regardless of what it may involve, that need is of the same importance as any other need involving some other state of affairs. Though it may be granted that one could have a hierarchic ordering of needs, the temporal differentiation does not determine that one need within the hierarchy is of greater value than another. In this respect, one might argue that the concept of need resembles the concept of possibility, in that the latter (unlike the concept of probability) also does not allow for degrees.

Thus, one can reason that if 'need' is the modality, which, like possibility, does not allow for degrees, then one cannot employ the methodology of economics to analyze preference within a philosophical context. For the latter deals with evaluating *degrees* of wants and desires in order to determine market demand, whereas needs cannot be analyzed in this manner. In this connection, Rescher refers to Alfred Marshall's classic work, *Principles of Economics*, so as to justify the general thesis that one can proceed by assigning numerical units of merit indirectly to various states of possible satisfaction. In fact, a number of instances can be cited where Marshall is arguing for the attaching of a unit of measure to manifestations of desire and want. Of the numerous examples, one could cite the following from the *Principles of Economics*: "....Utility is taken to be correlative to Desire or Want. It has been already argued that desires cannot be measured directly, but only indirectly by the outward phenomenon to which they give rise:... the price the person is willing to pay for the ... satisfaction of his desire...."[13] At this juncture it is important to note that Marshall's reference to desires and wants is one which eschews any connotation of inner mental states. Rather, Marshall sees desire, for example, as what is manifested in a supply and demand context. Hence, it is so to speak a derived sense of desire, based upon market demand. However, to read Marshall on this point is to find that there is no real loss in operating in this manner, since he sees desire as distinct from any ethical or prudential connotations. Moreover, his basic concept of "utility" is developed upon this highly sterilized conception of desire and want. In essence, then what Marshall is doing is interpreting

---

[13] Marshall, Alfred, *Principles of Economics*, 9th Edition, New York, The Macmillan Company, 1961, Chapter III, p. 92.

an overt form of human behavior of the market place in a sufficiently general way to cover some phenomena which his economic theory is designed to explain.

The issue of whether economics can be said to deal with desires or wants when referring to preferences is in itself worthy of further independent study. The resolution of this question seems to require the realization that there does not appear to be within economics any serious effort to distinguish preference from desire conceptually, which may be what leads Rescher into thinking that one can freely apply economic modes of analysis to philosophical investigations of preferences. Without entering beyond the immediate area of discussion, White's observations on the grammar of expressions of need call for a serious reexamination of how desire relates to preference—if at all. Thus, from the level of linguistic analysis, there seems to be genuine doubt whether economists and philosophers talk about the same thing when they speak about preferences.

II.

Rescher believes that the "formal characteristics" manifested in investigating discourse involving preferences are independent of the specific context one uses to illustrate merit assessments. Since his semantic approach is prefaced on the notion of "possible" world states, Rescher claims that certain formal relationships of preference can be generated without necessarily mentioning any specific instantiation. It is this claim which will be scrutinized next, especially since it utilizes the idea of the applicability of the mathematical model to the formalization of relationships of preference.

Looked at carefully, what is being done when numerical value is attached to a state description is an attempt to represent within a mathematical context some value judgments one is likely to make in ordinary discourse. However, the numerical units of merit one attaches to a state of affairs do not become criteria of individuation for these states of affairs, since the individuation has already occurred through the implicit evaluation performed by the person setting up the merit assessment of the states of affairs and their possible world distribution. Consequently, the issue emerges: whether the mathematical relationships Rescher sees as independent criteria for a logic of preference are really explicative of anything in such a logic, or whether these mathematical relations are actually inseparable from the calculus of preference itself.

By way of illustration, Rescher's "semantic" approach involves the setting forth of possible world situations, where the state of affairs of these worlds is taken to be intrinsically valuable and therefore meritorious. For example, one may have the following distribution:[14]

Possible Worlds		#(First-Order Preference) Value
w_1:	p&q	a
w_2:	p&-q	b
w_3:	-p&q	c
w_4:	-p&-q	d

The above shows that four world states can be specified, *given* the conjunction of two states of affairs expressed by 'p' and 'q'. Furthermore, it is *also* given that the "most" preferable world state is that where p and q obtain as a conjunction, and in the order where p is prior to q. Hence, the value expressed by 'a' is the highest value, and all the other possible world states receive values of proportionately lesser merit, e.g. b, c, and d.

Within the indicated distributions Rescher claims that one can observe "preference-tautologies", such as $pP^{\#}q \rightarrow -qP^{\#}-p$. Here '$P^{\#}$' is to be taken to express an extensional and ordering relation of first-order preference, and the entire expression can be read as: where p is preferred to q, then -q is preferred to -p. This is an "acceptable" preference-tautology *since* it "... goes over into an arithmetical truth."[15] Rescher illustrates this by showing how the above principle actually means:

$$\frac{a+b}{2} > \frac{a+c}{2} \rightarrow \frac{b+d}{2} > \frac{c+d}{2},$$

which reduces to the following arithmetical truth: $b > c \rightarrow b > c$. The quantitative "greater than" '$>$' is taken to replace '$P^{\#}$' of the natural language.[16]

It is important to observe that the preference-tautology is recognized as such only because it is seen to "yield" an arithmetical truth when it is considered in terms of its numerical value. The suggestion is that its real tautological nature is not evident without the introduction of the mathematical model. Rescher puts the matter more strongly by saying that in order for a proposition to be "accepted" as a preference-tautology, it "must" translate over to an arithmetical truth. The mathematical equivalence becomes the "criterion" for determining whether some proposition expresses a preference-tautology.

---

[14] Rescher, Nicholas, *The Logic of Decision and Action*, p. 44.
[15] ibid. p. 45.
[16] ibid. p. 45.

The immediate question seems to be: in what way is the mathematical element determining the tautological character of the proposition which expresses the preference-tautology. Rescher's so-called "criterion" for a preference-principle is that it must express an arithmetical truth. How is this arithmetical truth somehow "reflected by" the preference-principle?

In an ordinary language context one does not presuppose an arithmetical criterion when expressing a preference. Yet Rescher's approach requires that one interpret the relation of direct preference (expressed by 'p^*') in terms of greater numerical merit (expressed by '$>$'), and that this reveals the tautological character of the preference-principle. However, the mathematical model should be conceptually independent of the discovery of the preference-tautologies. For one must have the possible world permutations separate so as to determine the character of the preference-principle. Is it justified to say from Rescher's viewpoint that the preference-principles in some way reflect independent mathematical truths which emerge from the possible world distributions and their relative value assignment?

This question revolves around saying whether the entire set-up (the possible world states with their values) could be had at all without the evaluation process which sets down the values (a through d), superseding and determining the numerical context. Irrespective of the fact that possible world states do not refer to any actual states of affairs, the evaluational activity is distinctively an irreducible structural component which has direct bearing on the mathematical significance of the so-called preference-tautologies. Consequently, what kind of "criterion of acceptability" can these mathematical truths be, if *what* they are designed to examine is directly influencing their structuring?

The values expressed by 'a' through 'd' do not represent simply numerical quantities. Rather, they represent an ordering of presumed value relative to some possible state of affairs. Hence, these numerical assignments are set within a context of value, receiving their rationale within that context. This is an important point in that for Rescher, preference is conceived in terms of what is intrinsically valuable. Yet, one sees here how the very presupposition of intrinsic value is operating in the structuring of the mathematical model. For this reason it is difficult to see what the "criterion of acceptability" is designed to exemplify.

One has here again a Wittgensteinian point, namely, if something is going to count as a criterion, then there must be some way of "independently justifying" the *correct* application of the criterion to the thing studied.[17] How-

---

[17] Wittgenstein, Ludwig, *Philosophical Investigations*, G.E.M. Anscombe, translator, The Macmillan Company, 1970, pp. 59i - 60e.

ever, if the criterion itself is influenced by what is being studied, then no justification can be brought to bear showing that what is being analyzed is comparable to the independent standard, the mathematical units of merit in this case. Hence, it can be argued that the very operation of assigning numerical values to possible states of the world makes problematic the process of determining preference-tautologies.

The matter is different from, say, the case of the Pythagorean discovery that mathematical proportion governs musical harmony. The mathematical ratios of music are derived from *within* the musical composition itself. This is to say that the mathematical dimension of music is discovered within the compositional score as a natural characteristic of the music. There is no mathematical model which is independent of the music *to which* the score is compared and contrasted. In the case of Rescher's semantic analysis of preference, there is the suggestion that the mathematical dimension of the preference-tautologies is somehow the standard which the latter must translate into. Thus, the preference-principle must reflect the supposedly independent mathematical criterion. The problem here is specifically with the nature of this "reflecting," so-called. It is helpful to be reminded by the fact that Rescher, as well as all who employ the extensional approach to this issue, seek to present a formalization of preferences which is totally abstract and formal. Presumably, this would be a logical abstraction, in the sense of propositional logic. Yet, for Rescher, such formalization must also contain or reflect an arithmetical necessity, and how this is so is not adequately explained by him.

One also finds that Rescher's explanation of the relation of preferring is similarly presented within a highly mathematical setting. Considering again the preference-principle: $p P^* q \rightarrow -q P^* -p$, it is interesting to note that Rescher finds no difficulty in equating the meaning of the symbol for implication '$\rightarrow$' from a semantic context of preference to the meaning of '$\rightarrow$' in the supposed independent context of mathematics. The intracontextual use of this symbol underscores from another direction the point already made regarding the apparent non-independence of the mathematical context from the evaluative one of preference. Moreover, the symbol 'P^*' was seen to be expressive of an extensional ordering relation. By an "ordering" relation, Rescher means a relation that is transitive, asymmetric, and irreflexive.[18] There is little doubt that the 'P^*' is already couched in mathematical terms, and the relation the relation has to any ordinary sense of preference is difficult to decipher. Thus, the so-called

---

[18] Rescher, Nicholas, *The Logic of Decision and Action*, p. 42.

independence of the linguistic context of preference from that of mathematics cannot be supported by Rescher's use of his own symbol 'P*'. It is not instructive to claim that the above preference-principle, as well as others he alludes to, are tautologies *because they translate over* to arithmetical truths. The idea of translation presupposes *two separate* systems which, following Quine, are linked by some common index which enables the translator to convert expressions from one system into the other.[19] Here, however, both contexts of discourse (i.e. the one of preferences and that of mathematics) have been intermixed, and there is no index of translatability.

One can also realize the above inter-crossing of contexts in the incisive analysis of Rescher's work by Anthony Willing.[20] The latter's discussion deals with the way in which a counter-example suggested by Chisholm and Sosa to Rescher's preference-principle pPq → -qP-p is actually found to be consistent within Rescher's semantics. Briefly, Willing attempts to show that, by way of *reductio ad absurdum*, one can arrive at an explicit contradiction within a context of possible world distributions as set forth by Rescher, given the negation of the consequent of the Chisholm/Sosa counter example: pP*q → ((p&r)P*(q&r)&(p&-r)P*(q&-r)).

Willing prefaces the ensuing proof with the following important observations. First, he points out that there are two fundamental assumptions Rescher makes where he speaks of the derivation of valuations of possible worlds from 'raw' valuations of states of affairs. This is to say that Rescher's determination of the following possible world valuations:

Possible Worlds		#Valuation
w_1:	p&q	+4.5 units
w_2:	p&-q	+3.0 units
w_3:	-p&q	+0.5 units
w_4:	-p&q	−1.0 units

stem from intrinsic state of affairs valuations.

If it is the case that: , then the resultant utility-value is:

p	+4.0 units
-p	0.0 units
q	+0.5 units
-q	+1.0 units

---

[19] Quine, Willard van Orman, *Word and Object*, The M.I.T. Press, Cambridge Massachusetts, 1964, p. 26-28.

[20] Willing, Anthony, "A Note on Rescher's "Semantic Foundation for the Logic of Preference," *Theory and Decision*, Vol. 7, No. 3, July, 1976, p. 221.

Thus, possible world valuations are derived from the assigned ('raw') values of particular states of affairs. Moreover, one can determine the propositional valuations from the possible world valuations so that the former exhibit the following structure:

Proposition	#Value
p	+3.75 units
-p	-0.25 units
q	+2.50 units
-q	+1.00 units

One notices that in the above table, the propositional value of p(+3.75) is the value of w_1 plus the value of w_2, divided by 2, etc.[21]

The assumptions that Rescher makes in setting up these distributions are:[22] (1) the values of the possible worlds are a *function of* the sum of the raw values of the states of affairs which pertain to them; this, Willing calls the "Additive Assumption." In other terms, $\#(w_1) = V^r(p_1) + \ldots + V^r(p_n)$, etc. (2) the raw value of each proposition about a state of affairs retains this same truth value for every possible world within which it is true; this, Willing calls the "Retentive Assumption".

It is doubtful whether one need go fully into Willing's proof of his *reductio* argument, since its author concentrates on the purely formal deficiencies of Rescher's method. It *is* important, however, to discuss the two assumptions mentioned above and how they reflect the intercrossing of the mathematical context with the context of preferences.

Willing's point concerning the two assumptions mentioned above indicates the manifestation of two contexts, the mathematical and the linguistic (preferential) and how they actually seem never to be reconciled in Rescher's semantics of preference. A brief look at Willing's proof will serve to illustrate this point.

The counter-example given by Chisholm and Sosa, and said by Rescher to be 'unacceptable' as a preference-tautology, states the following:

$$pP^*q \rightarrow ((p\&r)P^*(q\&r)\&(p\&\text{-}r)P^*(q\&\text{-}r))$$

Willing offers the following possible world distribution, together with suggested value assignments:

---

[21] Rescher, Nicholas, *The Logic of Decision and Action*, p. 55.
[22] Willing, Anthony, "A Note on Rescher's "Semantic Foundation for the Logic of Preference," p. 224-225.

Possible Worlds				#-Values
W_1	p	q	r	2
W_2	p	q	-r	2
W_3	p	-q	r	4
W_4	p	-q	-r	4
W_5	-p	q	r	4
W_6	-p	q	-r	2
W_7	-p	-q	r	4
W_8	-p	-q	-r	4

Given the above table, one can show that the counter-example is false.

$$\text{For the value of } \#(p) = \frac{(2+2+4+4)}{4} = 3,$$

$$\text{and the value of } \#(q) = \frac{(2+2+4+2)}{4} = 2.5.$$

Hence, it follows that $\#(p)$ is of greater value than $\#(q)$, which means for Rescher that pP^*q. On the other hand, it is seen that

$$\#(p\&r) = \frac{(2+4)}{2} = 3 \quad \text{and} \quad \#(q\&r) = \frac{(2+4)}{2} = 3.$$

Thus, it is false that $(p\&r)P^*(q\&r)$ is true, since $(p\&r)$ is not greater than $(q\&r)$. Thus, if $(p\&r)P^*(q\&r)$ is false, then the *entire* consequent is false, being that it is a conjunction. This makes the whole counter-example false, given that its antecedent, pP^*q, is true and its consequent is false.

However, Rescher cannot claim victory for his refusal to accept the counter-example because of the above argumentation. For Willing proceeds to show that from another approach, the example is consistent within Rescher's semantics.

Given the antecedent as a premise in his *reductio* proof, Willing proceeds to argue as follows:

(1) pP^*q

(2) $-((p\&r)P^*(q\&r)\&(p\&-r)P^*(q\&-r))$, the negation of the consequent of the counter-example.

(3) $\#(p) > \#(q)$, definitional transformation of (1).

(4) $-(\#(p\&r) > \#(q\&r)\&\#(p\&-r) > \#(q\&-r))$, definitional transformation of (2).

(5) $(\#(q\&r) \geqslant \#(p\&r)) \vee (\#(q\&-r) \geqslant \#(p\&-r))$, definitional transformation of (4).

Willing proceeds to observe that if one can arrive at a contradiction between (3) and the first disjunct of (5), as well as a contradiction between (3) and the second disjunct of (5), then he would have proved the consistency of the counter-example in Rescher's semantics. For present purposes however, it will suffice to show how he proves the first contradiction.

(5a) $\#(q\&r) \geqslant \#(p\&r)$

(6) $\dfrac{\#(W_1) + \#(W_2) + \#(W_3) + \#(W_4)}{4} > \dfrac{\#(W_1) + \#(W_2) + \#(W_5) + \#(W_6)}{4}$,

the interpretation of the first disjunct of (5) according to the last table.

(7) $\dfrac{\#(W_1) + \#(W_5)}{2} \geqslant \dfrac{\#(W_1) + \#(W_3)}{2}$,

the interpretation of (5a) according to the last table.

(8) $\#(W_5) \geqslant \#(W_3)$, from (7) by implication.

(9) $\#(W_4) > \#(W_6)$, by implication of (6) together with (8).

Therefore,

(10) $(V^r(p)+V^r(-q)+V^r(-r)) > (V^r(-p)+V^r(q)+V^r(-r))$, applying the Additive Assumption on (9). This is to say that the value of the former part of the expression is collectively of greater value than the collective value of the latter part of the expression.

(11) $(V^r(p)+V^r(-q)) > (V^r(-p)+V^r(q))$, assuming the Retentive Assumption on (9). This means that the individual value of p, q, -p, and -q are such that p and q must be greater than -p and -q, and -r *simply cancels out on both sides of the expression*.

(12) $(V^r(p) = V^r(-q) + V^r(r)) > (V^r(-p) + V^r(q) + V^r(r))$, applying the Retentive Assumption on line (11). This is the brilliant move in Willing's proof. It shows that the value of line (11) is unaffected by the introduction of $V^r(r)$ on both sides of the expression. Yet by doing this, one is saying in effect that:

(13) $\#(W3) > \#(W5)$, the interpretation of (12) according to the last table above. By line (13), however, one has a contradiction with line (8). Hence, Willing has proven his *reductio ad absurdum*.[23]

Willing's proof reflects a tension between expressing value numerically, and expressing the explicit designation of a world state having value, such as $\#(W^3)$ and $\#(W^5)$. In this way, his proof focuses upon the weakness in Rescher's conception of the role of mathematics, functioning as a means of illus-

---

[23] ibid., pp. 226-229.

trating the rigor contained within a linguistic expression. For it can be seen in the proof Willing offers that the mathematical factor is not sensitive to the "meaning" of a possible world state, as composed of a number of states of affairs, in a particular sequence, and thus having a special identity. In this formal analysis, one sees that there is a gap between the mathematical context and the context of language, so that it really cannot be maintained that the former "elucidates" the latter.

The strength of this proof results from Willing's careful examination of the assumptions which are implicit in Rescher's semantics of preference. These assumptions are actually the bi-polar and antithetic aspects of Rescher's entire thesis. The former, the Additive Assumption, makes possible the mathematical component of his analysis, which is the standard by which one is to determine preference-tautologies. This is to say that the general value of possible world expressions which compose the content of preferences must reflect a summational character, since the values of their constituent states of affairs go into making their general intrinsic value. In short, the values of the individual possible worlds are formed as the aggregate value which results from adding the values of the particular states of affairs. Without the Additive Assumption, one would be unable to claim that a particular possible world is preferred over another different possible world. It is seen here that this first assumption actually makes possible the necessary numerical evaluation which forms the basis of preference-tautologies for Rescher. On the other hand, the Retentive Assumption insures the preservation of the individual identity or "uniqueness" of each possible world. For if the constitutive states of affairs expressed by 'p', 'q', 'r', etc. are not held as being always distinct, while constituting the molecular possible world state, i.e., they are not allowed to retain their particular identity, then there is no way of differentiating one possible world from any other. Hence, the Retentive Assumption is vital for preserving the separate character of the preference it addresses itself toward. This second assumption is useful in securing and clarifying the extremely important individuating aspects of expressions of preference, which is a way of accounting for the semantic component of the analysis. This is to say that the second assumption tries to accommodate the uniqueness of a possible world state, and in doing so reflects the special meaning of that particular state.

Basically, the difficulty with Rescher's entire thesis is that it is not clear how values are to be added, in a mathematical sense. The assumption is that numbers are going to stand for the values of possible world states. In saying so, however, one is contradicting the Retentive Assumption. For when numbers are added together, no restriction holds as to which numbers can or cannot be added together. Nor does it make any sense to say that a given aggregate sum will be distinct from an equal aggregate sum because it contains within it the addition of some particular number. Numbers have a sense which is determined by the system of mathematics itself, and if they are used to represent evaluative determinations, then they can only have a simple ordering capacity. Yet, Rescher claims that his logic of preference is not merely ordinal but evaluative as well. He begins by having the units of merit operate as numbers ordinarily do, and proceeds to generate the tautological character of preference-principles basically by way of allowing for the addition, subtraction, and division of numerical units of merit. In all of this, there is no discussion as to how these units of merit can function both as numbers and as indices of evaluative distinctions. What sense does it make to say that "units of merit" can be added, etc. while at the same time trying to claim that each value is distinct, unique, and "intrinsic", as the Retentive Assumption insures?

One may also argue that Rescher's Additive Assumption appears to incur the fallacy of composition in the attributive sense. For Rescher is seen as saying that, because certain states of affairs are found to be desirable and are thus given x-number units of merit, therefore the value of the whole possible world which is composed of these states of affairs is the aggregate value of all these units of merit, and thus this aggregate value is greater (preferred) than any other value of a possible world whose constitutive states of affairs have less numerical value. Of course, it does not follow that, because an individual finds desirable certain states of affairs which go into making a whole possible world, he will find that whole possible world desirable as well. Yet, the numerical method which Rescher uses reduces all values to the common denominator of number, and this leads one to reason that the more desirable state is the one whose aggregate reflects the highest units of merit.

Willing does not consider the full ramifications of his discovery beyond that of showing the "formal acceptability" of the Chisholm/Sosa counterexample. His work, when set in a broader perspective, relates to a key difficulty with Rescher's analysis of preference. Willing's concentration on the two seemingly irreducible assumptions in Rescher's semantic analysis illustrates graphically that the mathematical and the linguistic contexts are not reconciled. This comes out exactly at the point where one must decide when to attach either a numerical value to an entire molecular possible world state or to endeavor to provide an effective means of differentiating one possible world from another. The logical incongruity Willing has brought out is in essence a reflection of Rescher's apparently unconscious blending of these two indissolubly distinct realms of discourse.

Though Willing's criticism of Rescher's semantic theory of differential preferences seems cogent, it must not be taken out of context, to the extent of claiming thereby the demise of Rescher's entire effort. Willing has focused upon an incongruent detail in what Rescher presents, and as such it deserves mentioning in terms of its *possible* ramifications for the Retentative and Additive Assumptions, which are important for Rescher's semantic theory. At best, however, these are suggested implications of Willing's analysis, and must be seen as such. They point to an incoherence in Rescher's #-semantics, but again much remains intact as an interesting and innovative approach to the formalization of preference expressions. Moreover, though it may be pressed that Rescher erred in thinking that he could generate a counter-example to the Chisholm and Sosa approach by using "raw" values of preferability, Willing's criticism is still one of detail and it should not effect the entire picture of what Rescher has to offer. Nonetheless, it deserves consideration as a segment of the literature in this area, and as such it is presented here. The real issue which shadows Rescher's approach, and the one which is foremost in the design of this chapter, deals with the success Rescher's approach has in capturing the sense of the logic of discourse concerning preferences as expressed in ordinary discourse. It is this issue especially which the forthcoming discussion concentrates upon.

III.

Rescher sees his semantic approach as a "superior" means of analyzing preferential discourse, in contrast to axiomatization, with its reliance on "faulty intuition".[24] He grounds his claim on the belief that he has arrived at a means by which preference-principles can be determined according to precise mathematical criteria. In fact, he lists the different philosophers who have tried to axiomatize preference, and proceeds to show their confusion in how they disagree on which preference-principles are intuitively valid. Though Rescher's research on this point is sound, with a few exceptions, one must inquire whether the alternative he offers is better as far as disentangling the notion of preference is concerned.[25] The preceding section illustrated that Rescher has not succeeded in convincingly securing his objective. The ques-

---

[24] Rescher, Nicholas, *The Logic of Decision and Action*, p. 54.
[25] ibid. p. 54.

tion remains whether the kind of precise criteria Rescher is searching for are at all applicable to his conception of the analysis of discourse involving preferences in terms of possible-world states.

It has been seen that Rescher centers his analysis around expressing mathematically the evaluative aspect of preference. Preference is considered by him in an intrinsic sense, whose value is translated into and clearly expressible in terms of precise mathematical units. However, at the outset one is faced with the assumption that the value of the preference is some sort of object *to which* the merit - as a number - *is attached*. This makes *the* value of the preference manifest itself as an entity, somehow existing *independently* of the expressed preference. Yet, it would seem more correct to say that the value of an expressed preference emerges from within an interaction of an individual with his environment, and that apart from this interaction, *the* value of a preference as an independent intrinsic entity makes no sense. This would be consistent with the view expressed above. Namely, being that "need" is the modality which underlies the notion of preference, the latter can only be understood in extensional terms, — i.e. in terms which involve relevant instrumentality.

It was seen in Section II that Rescher speaks of the "propositional valuations" emerging from possible-world valuations. The problem with this way of handling the formalization of preferences seems to be analogous to the difficulties Alan R. White discusses in his perceptive book, *Truth*.[26] White points out that it often happens that philosophers speak of words or sentences *bearing* their "meaning" or "proposition". Thus, there is a tendency to objectify the sense of the uttered expression, and then to say of this objective meaning or "proposition" that it is true or false, accurate or exaggerated, etc. White continues:[27]

> "...What is said is embodied in and has no existence separate from the various media in which it is said any more than a shape, which may be common to many objects, exists separately from the objects which have this shape. This is why what is said can occur wherever and whenever the appropriate words occur, e.g. in the Nicene Creed or in the pulpit on Easter Sunday. This is why to read, take down, print, preserve, alter, or destroy what is said is to read, take down, print, preserve, alter, or destroy what is uttered or written. The introduction of a special name for what is said in an

---

[26] White, Alan R., *Truth*, Anchor Books, New York, 1970.
[27] ibid., p. 16.

utterance, e.g. "proposition", "statement", or "judgment", *leads us to overlook these indissoluble connections between what is said and the medium in which it is said.* (my italics)"

Though White gears his remarks toward the analysis of the notion of truth, much of what he says can be used to scrutinize the notion of preference as Rescher conceives of it. For White is saying that one cannot separate the meaning of what is said from what is said as a verbal utterance. Yet Rescher permits this very same separation where he talks of the intrinsic value of the preference as *something to which* a unit of merit is to be applied. In divorcing the value of a preference from the expressed preference, one is caught talking about an entity which has no meaning apart from the context of the preference itself. What is *the intrinsic value* of the preference? Apart from the particular preference itself, it has no sense of its own; it is vacuous!

The above point can be brought out quite clearly where one recalls the discussion in Section I, which differentiated expression of preference from expressions of desire and want. There it was seen that because need plays an important part in the determination of preferences, one can be preferring things which are not desired, simply because they are needed. Thus, the values of preferences are "indissolubly" connected with the world surrounding the person doing the preferring. Apart from this pragmatic relation between language-user and the extension of the utterance he uses to express what he prefers, the idea of intrinsic preferential value existing as somehow separate from the context of the preference itself simply fails to signify anything. Thus, the ostensive conditions which constitute or set the stage for the preference must be taken into account, so as to secure a cogent explanation of a preference's value. Where this is not done, however, then an internal inconsistency emerges when trying to define the concept of the value of a preference. For the extension of this concept requires reference to the conditions which constitute the need underlying the preference. Yet, it is seen that Rescher employs a view of preferential-value which entitizes it into a self-contained and independent thing. Continuing from this, he considers such value as a *type* of entity, whereas evidently it can only be spoken of in terms of it being a *token* of the interaction of a number of conditions. Hence, there is an inconsistency between the objectively determinable extension of the concept of a preferential-value, and Rescher's conception of its intentional character.

It can be argued, however, that Rescher chooses the semantic mode of analysis as a way of more suitably accommodating preferential-value within a possible-world mode of analysis. This is to say that where preferential-value is seen as a context-free individual, then the variability of the value of the preference can be conveniently expressed as represented by a variety of possible-world states. Hence, *the value* of the preference, as an individual to which a unit of merit is attached, is taken as functioning as a variable within a possible-world mode of analysis.

Thus, one is led to the broader question of whether Rescher can credibly formalize the semantics of preference by means of a possible-world mode of analysis, while also conceiving of preferential-value in wholly intrinsic terms. In essence, it can be said that Rescher's aim is to elucidate formally the subjunctive *intention* of preference by means of a possible-world analysis. The approach itself is not being questioned. This is to say that throughout Rescher's presentation, no difficulty was raised concerning the ontological status of possible-world states, and why this method of analysis requires that it be entered into only when one by-passes certain critical philosophical problems. Throughout the prior section, possible-world states were allowed to function exactly as they were intended by Rescher, i.e. as individuals in a propositional calculus. This was done for the sake of allowing Rescher's position to unfold.

Nevertheless, when this method of analysis is applied to preferences, serious difficulties are encountered relative to its suitability as an analytical tool. For example, in section II, tension was seen to be manifested between the mathematical formalisms Rescher tries to derive and the linguistic character of preferential-discourse, considered in connection with its distinct evaluative component. Such difficulties may be taken as suggesting that, due to the fact that preferences can be interpreted in terms of subjunctive (future perfect) conditionals, on a deeper level they do not lend themselves to a possible-world type of analysis. No doubt Rescher is aware that such analyses have been employed as a means of getting around some of the problems raised by R. Chisholm and N. Goodman on the analysis of conditionals in the context of law-like explanations in science. However, though Rescher's presentation of both first-order and differential-preference is presented in subjunctive terms, this by itself *may not* justify using the possible-world method of analysis when dealing with preferences. It is important to note that where this method has been used to elucidate the signification of scientific laws, issues of evaluation do not play a direct role.

Hence, one comes to the problem of the suitability of possible-world state analyses for the formalization of expressions of preference. A related discussion becomes very sharply defined in a recent talk by Professor Stalnaker, entitled "Formal Semantics and Philosophical Problems"[28] In outlining the difficulties which a review of the history of possible-world state analyses reveals, he illustrates with admirable clarity the shortcomings of some of these efforts bearing directly on the matter at hand is the thesis he develops concerning the differentiation between the indicative-conditional and the subjunctive-conditional. The first is illustrated by the example: "If Oswald did not shoot Kennedy, then somebody else did." The second is illustrated by the example: "If Oswald had not shot Kennedy, somebody else would have." Stalnaker argues that the former, the indicative-conditional, can be given a cogent interpretation in terms of possible-world state descriptions. However, the latter subjunctive-conditional, as yet defies formalization in terms of possible-world states. The reason for this is that the causal connectedness between the antecedent and the consequent of such conditionals is of a much more illusive nature than one finds in the indicative-conditional.[29]

Stalnaker points out that it is difficult to specify what states of affairs will satisfy the conditions expressed by words such as: "had not" and "would have". This is because such terms refer to no one determinable state, but to an indefinite contingency of possible-world states. Moreover, any causal connectedness between these conditional states within the subjunctive mode virtually defies pinning down.

He goes on to refine this point by noting that, in the case of an indicative-conditional such as: "If Oswald did not shoot Kennedy, somebody else did.", one has the situation where if the antecedent turns out to be false, then the consequent is still true, since the facts confirm the case that President Kennedy was indeed shot. Hence, the context in which the indicative-conditional is uttered functions closely with the sense of the expressed indicative-conditional. This same close relationship between context and expressed conditional is not evident where one deals with subjunctive-conditionals. Again, considering the case he gives: "If Oswald had not shot Kennedy, somebody else would have.", one sees that in the subjunctive mood the expressed conditional introduces a wider variety of possibilities into the meaning. This is to say that the subjunctive-conditional perhaps implies that had Kennedy not been shot, his performance in office would have, could have (?), induced his assassination. This, together with many other nuances of meaning, appears to be suggested by the subjunctive. Stalnaker observes that from these two examples it becomes clear that the indicative-conditional imposes a greater constraint on its

---

[28] Stalnaker, Robert. "Formal Semantics and Philosophical Problems," Paper presented at the American Philosophical Association Meeting, Dec. 1979, typed manuscript, p. 6.

[29] ibid., pp. 6-7.

context than the subjunctive-conditional. Consequently, it is more reasonable to advocate the application of a possible-world state analysis to indicative-conditionals than to subjunctive-conditionals. The subjunctive-conditional assumes the suspension of the background of presupposition which the indicative-conditional does not. For this reason one would expect the subjunctive-conditional to have a richer variety of connotation, which the indicative-conditional does not possess. On the other hand, the indicative-conditional expresses a clearer causal connectedness between antecedent and consequent than one finds in the subjunctive-conditional. The causal connectedness of the former is again a reflection of the contextual constraint which the indicative-conditional demands, and which the subjunctive suspends.[30]

The relation of all this to the interpretation of expressions of preference, considered in terms of possible-world analyses, turns upon the fact that Rescher construes expressions of preference, either first-order preference or differential-preference, in terms of subjunctive-conditionals. This makes the application of possible-world analyses to expressions of preference difficult to accept in view of Professor Stalnaker's observations. One recalls how Rescher expresses first-order or direct-preference as that desirable state one would prefer *should* some state of affairs come about. Moreover, he considers differential-preference as the one state of affairs between two competing possible-world states which *would* be desirable. Thus, in the case of both differential and direct-preference there is a presupposition of the subjunctive mood, though usually with the disguise of the gerundive, such that one has expressions such as: "p's *being* ... (the case) ..." and the "... *happening* of p ..."[31] If one considers the way Rescher explicates his preference-principles, one finds that the gerundive locution hides the implied "... if p *were* preferred to q, then it *would be* the case that such and such *would be* preferred ..." Quite clearly, in light of Stalnaker's insights concerning the contextual suspension which the subjunctive involves, the issue directly ahead is whether a possible-world type of analysis can be used to elucidate the semantic depth of preferential discourse.

Can one apply a possible-world state analysis to explicate the semantics of preference, rendered in the manner of subjunctive-conditionals? This question would not merit a blanket negative from the standpoint of Stalnaker's analysis. Rather, as he points out, there are some serious and unresolved problems which those who have engaged in these types of analyses simply have

---

[30] ibid., pp. 7-8.
[31] Rescher, Nicholas, *The Logic of Decision and Action*, p. 40-41.

not come to grips with. Apart from the issue of the suspension of the background of presupposition which the subjunctive-conditional apparently necessitates, there is the question of explaining what sort of relationship, if any, holds between that which is preferred, and the consequence it entails. If one looks closely at an expression of preference, especially when taken in the form of a principle, such as: p P q → -q P -p, one finds that it expresses a generalization concerning that *where* p is preferred to q, then it *would* be the case that one *would* prefer -q to -p. Rescher's position here is that to prefer p to q "means" that one *would* prefer -q to -p.[32] Yet, what is the sense of "means" here?

One possible interpretation would be that if one knows of some agent, say X, who prefers p to q, then on some comparable occasion it would be expected that X would prefer -q to -p. Hence, his preference of -q to -p seems to be conditioned by his preferring p to q. That this is a logical implication seems dubious. Rather. here one is evidently dealing with an "empirical relatedness," which achieves expression linguistically. Rescher, of course, argues that such principles express a mathematical truth, which is the result of certain regularities between numerically-indexed possible-world states. Apart from the framework, of his analysis, it can be seen that expressions of preference-principles, taken as expressions of need with respect to a possible situation, are somehow tied to a context of action, involving an agent and his immediate or proximate environment. Thus, to prefer p to q does entail that the consequences of preferring -q to -p are the same as in the antecedent. In essence, is it not the case that to choose p to q is the *same* thing in terms of effects as that of choosing -q to -p? For it is the *effect* of preferring which is the same in both cases.

Yet, if the above preference-principle expresses in some specifiable sense the sameness of effect between the antecedent-preference and the consequent one, how is this sameness to be explained by means of a possible-world mode of analysis? Is it sufficient simply to have reference only to the numerical merit one attaches to world states, and to argue that both preferential expressions are the same because they exhibit a numerical identity? The problem is more complex, however, due to the fact mentioned earlier, namely: that the subjunctive-conditional always tends to suspend the contextual background, and, hence, it is not quite clear how any manner of numerical indexing can serve as a means of pointing out a sameness relation between antecedent and consequent,

---

[32] ibid., p. 42 (One must keep in mind that for Rescher 'P' is a symbol which expresses an ordering relation which means that '→' must also express the manifest ordering of that relation.)

which constitute the expression of a preference-principle. The question then turns squarely on the issue of the explanatory power of the possible-world mode of analysis, in light of the subtlety of preference-principles, when expressed in terms of the subjunctive-indicative.

Virtually the same difficulty is recognized by Stalnaker where he reviews the efforts of those who have endeavored to explain formally the nature of contrafactuals, who also used possible-world modes of analysis to account for the "connections and dependencies among facts: mysterious, unobservable ties that relate observable events of the world." In this enterprise, Stalnaker reminds the reader of C.I. Lewis' efforts at finding some criterion of sameness by which to relate distinct possible-worlds, while also trying to avoid the circularity of assuming the very connections one is trying to explicate.[33]

Interestingly, one finds in the case of the Rescher's setting forth of preference-principles a similar attempt at capturing within a framework of possible-world states the necessary implication of the act of preferring, i.e. what the '→' sign is interpreted as meaning. Yet he, like Lewis, does not consider what the criteria of sameness are so as to bring off this characterization in a credible fashion. Somehow the idea prevalent in Rescher's work is that preference-principles must express transitive and irreflexive relations between the antecedent-preference expression (i.e., pPq) and the consequent preference expression (i.e., -qP-p). Though it is not disputed that a preference-principle of this kind may exhibit transitivity and irreflexivity, to understand and accept this in Rescher's semantic analysis, some explanation of what it is for a possible-world state to be the same as another world-state must be given first. Simply to refer to to an identical numerical sum for world-states hardly captures the complexity of the issue at hand. Moreover, this complexity is deepened by the apparent role of time in the expression of preference-principles. For it is seen that such a principle is saying: if p is preferred to q at time t, then -q is preferred is preferred to -p at time t_1. The implied sense here is that if the former were preferred, then the latter would be preferred as well. Unless a principle is couched in these subjunctive terms, its instrumentality is seriously undermined. However, how is its time element to be captured within the framework of a possible-world mode of analysis? Time is a vital aspect of the nuance of meaning of the expressed meaning; yet this element seems to be beyond what the possible-world state can characterize. Of equal interest here is the role of intention in preference. Surely in preferring one is choosing between possible alternatives. Thus, there is a conscious intending to act

---

[33] Stalnaker, Robert., pp. 11-13.

which forms part of the meaning of preference, and which again seems remote or beyond access of a possible-world mode of analysis at this time.

The difficulties which are being pointed out here here serve to suggest that there is a shortcoming in the method of analysis proposed by Professor Rescher. In essence, it is the difficulty of describing in terms of possible-world states analysis the subtleties of language which discourse about preference seems to presuppose. Thus, it is not easy to accept the view proposed by Rescher at the end of his paper, to the effect that the semantic analysis he is proposing is superior to the axiomatic attempts initiated by others in this field. Rescher's point that the axiomatizers have not achieved a consensus as to what the principles of preference are, and thus their efforts are in hopeless disarray is perhaps too premature. One could always counter by saying that Rescher's effort does not consider the full range of a preference's reference when he attempts to present its logic, that is, he does not refer to the role of agent, time, object, action, intention, etc. Simply to attach a number to a possible-world state is not to achieve precision in the analysis of this semantic entity. Though the possible-world mode of analysis may provide a future means of expressing linguistic subtleties, at the present time much remains unexplained relative to the way mathematical structuring can tell us anything about the precision of language.

Throughout this analysis of Rescher's work, the attempt was made to see preference, or rather to interpret it, in terms of needs within a specifiable environment. In this way, it was hoped that the full richness of the notion of preference could be understood, and that Perhaps it could be shown that Rescher's work only dealt with a limited aspect of preference. At this juncture, it is perhaps best to recall Georg Henrik von Wright's insights concerning his three-part characterization of the range of moral analyses, namely, *Deontology*, *Axiology*, and *Philosophical Anthropology*. The latter he envisions as the study of concepts such as "needs and wants, decision and choice, motive, end and action."[34] Von Wright continues by observing that "moral philosophy is a special study of concepts of all three groups." Here it is important to observe that von Wright clearly states that moral philosophy must proceed from a careful study of these three groups, meaning that the clarification of ideas in this area forms a kind of prerequisite for understanding the moral concepts which reside at a higher level of theoretical abstraction. This is in essence the line of thinking which has been followed in the criticism of Rescher's viewpoint.[35]

---

[34] Von Wright, Georg Henrik, *The Logic of Preference*, Edinburgh University Press, 1963, p. 7.

[35] ibid., p. 7.

Basically, needs require to be considered together with decisions, choices, motives and actions - prior to proffering any conception of preference. To use von Wright's terminology, *Philosophical Anthropology* must be set down first, and then the concept of preference can be set forth and understood in terms of the richness of its implications and logic. Von Wright's observations are geared toward the conception of a logic of preference which has applicability in enriching the logic of moral discourse, and thus calls for a conception of preference which is more complete in its scope. Professor Rescher, as it was seen, models his logic of preference on a conceptualization found in economics, and as such its applicability to philosophical investigations may be severely limited.

In retrospect, it must be said that Rescher's attempt is a significant step in its time. It represents a new direction, in that, as he mentions at the very beginning of his paper, his aim is to bridge the gap between the "mathematico-economic" approach at developing a logic of preference, and the "logico-philosophical" one. Yet as such, it *may be* hampered by the very methodology he seeks to apply to the broader area of philosophical analysis. This is to say that, whereas the possible-world type of analysis may have applicability in the relatively defined area of economic decision-making, its applicability in the wider context of philosophical discussion may not be as smooth as Professor Rescher apparently assumes it to be. This is not to say that his approach has no merit in itself, and that it is therefore impossible to achieve the kind of mathematical rigor Rescher demands as a criterion for determining preference-principles. Rather the point is that there are unresolved problems in the way of accepting the possible-world mode of analysis, as Professor Stalnaker astutely points out. These difficulties prevent acceptance of his overall thesis that the approach he is pursuing is clearly superior to that which endeavors to axiomatize preference principles in terms which seem intuitively certain.

Chapter 4.

Richard C. Jeffrey's Logic of First and Higher-Order Preferences

In his article "Preferences Among Preferences" Richard C. Jeffrey endeavors to present a logic of preferences having successive orders. He claims that the formal analysis of complex implicational relations involving preferences (independently of how "first-order" preferences are interpreted) is possible through the use of a new preference connective belonging to the same syntactic category as C. I. Lewis' symbol '⥽', for strict implication. This connective is taken as providing a means for clearly expressing the intricate "modal involvement" which discourse concerning successive orders of preference comes to presuppose. Jeffrey's innovation in part is to treat the mutually self-excluding constituents of preferences as sentential *relata*, which when related by the preference connective 'Pref', are found to convey complex molecular propositions.[1]

The significance of Jeffrey's paper in 1974 can best be appreciated by considering it in relation to his book *The Logic of Decision*. What he has to say about the formal properties of successive orders of preferences is taken by him as an extension of his insights concerning the formal properties of first-order preferences in his earlier work. Indeed, one of the key goals in his later article centers around the question of how to deal with preferences of whatever order in a way which is consistent with his analysis in *The Logic of Decision*.[2]

Consequently, attention will be directed first toward his perception of a logic of preference in his earlier effort. This will constitute a foundation, so to speak, from which to consider what he terms "preferences among preferences." The second phase of the exegesis will involve his 1974 paper, which will be treated from a number of interrelated perspectives. Consideration here will be first to the discussion of the locutionary status of the sentential expressions, i.e., *relata*, which flank the 'Pref' connective, and to the analysis of the

---
[1] Jeffrey, Richard C., "Preference Among Preferences", *Journal of Philosophy*, Vol. LXXI, 13, July, 1974, p. 378
[2] Jeffrey, Richard C., *The Logic of Decision*, The University of Chicago Press, 1983.

entire preference expressions as conceived by Jeffrey. Secondly, the function of the 'Pref' connective will be investigated, in the way in which it differs from being interpreted as a two-place predicate expression. Third, attention will be given to certain cognitive assumptions regarding his analysis of preference. Finally, consideration will be directed to what it means to claim that higher-order preferences are to be seen as "preferences among preferences."

The four points outlined above constitute an integrated field of inquiry. For example, the investigation into the referential status of the *relata* which flank the preference connective leads in turn to questions concerning the tenability of employing a binary connective to evolve a formal language, designed to describe the modal properties of successive orders of preferences. This in its own right incurs issues involving the cognitive presuppositions which are taken as constituting "the knowing" of whether something is or is not preferred. Finally, considerations dealing with the latter entail one's focusing upon the very significance of talk about higher-order preferences.

At the outset, Jeffrey employs the technique of applying a relatively simple probability analysis theory to elucidate the sense of discourse concerning first-order preferences, and of their proper hierarchic ordering. Throughout his work in *The Logic of Decision*, he is found to allude to unanalyzed "propositions" that express preferences, and how these proposition come to constitute the *relata* that make-up the preference relations. Thus, he is keenly interested in the "human sense" of his proposed logic of preference, so that its mathematical character is not out of touch with the readily determinable sense of ordinary discourse concerning preferences. One also sees with Jeffrey's effort a maturity in the development of this area of linguistic analysis, where the object of investigation, namely spoken discourse, is taken as an important facet of the inquiry, along with the analytical tools of probability theory themselves.

In an interesting way Jeffrey's work compliments a direction of investigation initiated by Ramsey. This is to say that if an examination of Jeffrey's inquiry makes any pretense at being complete, then it must face the especially difficult task of noting the many differences between Ramsey and Jeffrey with respect to their views on the nature of inductive inference. As it was seen in the presentation of his views on the proper formalization of preference, Ramsey rejects Keynes' conception of inductive inference as basically that of justifying one's conclusion by appealing to an accumulation of known propositions. Ramsey argues that one's degree of belief in a proposition is more accurately determined by considering the notion of utility in relation to objectively determinable choice. Jeffrey, however, in more recent statements on this matter, seems to find it necessary for investigators to return to something of Keynes' view on inductive inference with respect to the issue of establishing one's degree of belief. A discussion of these differences, as they pertain to the notion of induction, and specifically as this plays a role in their respective conceptions of a logic of preference, is an unavoidable requirement of a critical look at Jeffrey's position in this area.

Section I. *Jeffrey's Logic of First-Order Preferences*

In *The Logic of Decision*, Jeffrey specifies that preference is reciprocally defined in terms of the notion of subjective probability and desirability. The former is articulated according to three basic propositions:

  A: it will rain tomorrow.
  B: it will snow tomorrow.
  C: it will blow tomorrow.

To the above Jeffrey assigns a sequence of eight positive numbers, for each of the eight possible cases in which these three propositions are true. For example, one case may be where all three propositions are true, e.g., ttt, or where A and B are true, but C is false, etc. On the other hand, desirability is represented by a sequence of eight numbers, which may be either positive, negative, or zero, without restriction, and which are attributed to each of the eight permutations of A, B, and C.[3]

From the above Jeffrey proceeds to define the "probability" of a proposition as "*simply the sum of the probabilities of the cases in which it would be true.*" In the case of compound propositions, as for example A v B, probability is determined by adding the numerical values of these cases in which both of the propositions are true. Similarly, $\bar{C}$'s probability is determined by computing the sum of the values wherein C is false.[4]

The only necessary propositions in Jeffrey's logic are of the form: A v $\bar{A}$, B v $\bar{B}$, and C v $\bar{C}$. For such compound propositions which are true in all cases he has the symbol T. In a similar manner, F is the symbol for propositions which are false in every possible case; the latter are the only propositions in Jeffrey's logic having 0-probability.

---

[3] ibid., p. 74.
[4] ibid., p. 75.

The desirability of a proposition is defined in an equally direct manner. Jeffrey says: "... *the desirability of a proposition is a weighted average of desirabilities of the cases in which it is true. where the weights are proportional to the probabilities of the cases.*"[5] The computation of desirability can be characterized by means of the equation: (desirability of each elemental proposition in a compound) X (their probability) =

<u>(the sum of the probabilities of these elemental propositions when true)</u>
(the compound proposition's desirability)

Having a means of deriving the numerically expressed desirability of a proposition, Jeffrey proceeds to present his rendition of *preference-ranking*. Simply put, a preference-ranking is the ordering of desirable propositions, according to the formula presented above. These four axioms are said by Jeffrey to yield precisely the same result as the rule stated earlier regarding the computation of desirabilities as the weighted averages of the desirabilities of the cases in which they are true.[6]

It is interesting to note R. M. Martin's general observations concerning a possible interpretation of Jeffrey's logic of preference up to this point. Martin claims that Jeffrey is dealing with two traditional senses of preference, namely *evaluative* preference, and *cognitive* preference. However, his innovation is in dealing with both of these senses within a unified mathematical theory, in that Jeffrey is attributing to propositions both utility or *subjective* desirability and *subjective* probability.[7] Much earlier, of course, Schick recognizes this same innovative aspect of Jeffrey's analysis in his review of the 1965 edition of *The Logic of Decision*, and Schick goes on to note how Jeffrey proceeds to introduce an additional change with regard to how one computes the desirability of actions. For Jeffrey, determining the desirability of action requires determining the desirability of the proposition which represents Jeffrey, determining the desirability of action requires determining the desirability of the proposition which represents the action. Thus, one also has the means of expanding the analysis by introducing the disjunction of the propositions expressing the performance of the action desired under any exhaustive set of incompatible conditions

---

[5] ibid., p. 78.

[6] ibid., p. 81.

[7] Martin, Richard M., *Language & Art, Essays in Honor of Nelson Good*, The Bobbs-Merrill Company, Inc., R. Rudner and I. Scheffler, eds., Indianapolis, pp.245-246.

(which are also represented as propositions). In this, Jeffrey reinterprets the Bayesian principle requiring the performance of an action which maximizes the realization of what is desired. In Jeffrey's rendition, as illustrated by the equation above, one has the introduction of the "conditions" of probability, taken as propositions, which are also considered in relation to the action to be taken. What is important to note here is that conditions considered in this manner are not probabilistically independent from the performance of the actions. This provides for a much more realistic or as F. Schick observes: "more serviceable," interpretation of the complex cases where preferences are operative. For example, the same results would ensue whether one were employing the Bayesian or the Jeffrean rendition, where the case involves say the action of wearing one's raincoat; given the condition that it will either rain or not rain. However, the results would not be the same where the case is that of one's abstaining from smoking, given the condition that one will live less than 65 years, and the condition that one live 65 years or more. In the Jeffrean method of computing probabilities, one can contend more easily with the complex situation where one's not smoking will lower the probability of one's dying before the age of 65. This kind of depth is not possible within the Bayesian framework of analysis, since it cannot deal with the disjunction of incompatible propositions.[8]

Having noted the above, it is equally important to bring out how Jeffrey's theory is also noncausal. In fact, he specifically denies the suggestion that there is a causal notion operating within his logic of preference. Rather, he views preference-ordering purely in light of propositions understood by some agent at a particular time, and the evaluation of the desirability of an outcome, also expressed as a proposition. There is no independent causal agent which would change the outcome of a probability assignment due to the outcome of some specific proposition. In this, Jeffrey's departure from Frank Ramsey is quite striking, in that the latter trades heavily with the *effects* upon outcomes by one's gamble to act in one way rather than another. This is seen throughout Ramsey's analysis of the notion of subjective probability, and especially where he explains *reasoned indifference* between options as the case where the agent knows that either one of two opposing modes of action would be noninfluential upon the outcome in some particular case. Jeffrey, however, having a clearer ontology than Ramsey's, in that both probability and desirability is at-

---

[8] Schick, Frederick, "The Logic of Decision Richard C. Jeffrey," (review), *Journal of Philosophy*, 64, 1967, pp. 396-398.

tributed to propositions, presents a less restrictive logic of preference. For it is not subject to the unneatness of the possible effects of gambles. Rather, as has been noted, and as will be developed further on, Jeffrey determines the probability function for propositions directly from desirability functions, within a coherent preference-ranking.[9]

However, rather than pursue more advanced aspects of Jeffrey's work at this juncture, it is instructive first to scrutinize his remarks concerning the nature of propositions, since these play a fundamental role in his theory. In reviewing *The Logic of Decision*, 1965 edition, R. M. Martin proceeds to excuse Jeffrey's allusion to "unanalyzed" propositions, and endeavors to interpret certain key ideas in Jeffrey so that they reflect his own insights into the formalization of preferences. For example, where one has in Jeffrey the expression for the degree of greatest probability, i.e., '$prob(a,X,t) > prob(b,X,t)$', or cognitive preference, one can introduce extensional pragmatics and interpret this as: 'X CogPrfr a,b,t' meaning that X cognitively prefers a to b at time t. Similarly, one could take what Jeffrey says about the desirability of a over b, or $des(a,X,t) > des(b,X,t)$, as X ValPrfr a,b,t, expressing how X evaluatively prefers a over b at time t.[10]

However, in his interpretation of Jeffrey, Martin is perhaps too quick to by-pass the issue of Jeffrey's allusion to "propositions," as not germane to his immediate concern of seeing how Jeffrey's work could be accommodated within his own method of analysis. Jeffrey devotes more attention to the notion of proposition in the 1983 edition of his book, and it is very important to consider carefully his views on this crucial notion, partly because they dramatize more so than in any previous attempt a serious concern with the linguistic component, as it reflects one's articulation of a logic of preference. It is no longer the case that one can proceed with a logic of preference without being greatly concerned with what Jeffrey terms "ordinary talk" about preference. This is a distinct departure from efforts in the late fifties and sixties where the role of language as a way of understanding the requirements for a logic of preference were only dimly perceived at best. However, it is also important to point out that though Jeffrey demonstrates a keen sensitivity to linguistic usage, he underscores the point that his analysis is not an exercise in linguistics.

---

[9] Jeffrey, Richard C., *The Logic of Decision*, pp. 59-60.
[10] Martin, Richard M., *Language & Art*, pp. 245-247.

Moreover, the notion of proposition deserves careful attention also because it has been seen to be the basis of his unified theory of preference.[11]

In the fourth chapter of *The Logic of Decision*, Jeffrey observes that there is a difference in attitudes between believing the proposition that it will rain tomorrow, and in desiring that it will rain tomorrow. He points out that one could argue that though one ordinarily accepts that it is things which are desired, it is nonetheless possible to *"suppose that various corresponding propositions are the actual objects of the attitude. ..."* To desire something, e.g., the love of a good woman, is to desire that the proposition concerning the good woman's love toward us "hold (= obtain = be the case = be true)." Interestingly, Jeffrey claims that what enables one to make the transition between things and propositions within the context of desiring is the "flexibility" of the notion of having, which in Jeffrey's view is strongly related to "desiring to have." Consequently, "to desire x is to desire that one have x," and thus what is desired is a proposition taken as true. In this connection, Jeffrey notes that he is not interested in reforming ordinary discourse, but rather in interpreting it. Significantly, Jeffrey refers to propositions without considering the issues which this concept has precipitated in the recent literature. For example, his remarks concerning the linguistic status of propositions touch only lightly on the possible distinctions which could be made among propositions, sentences, assertions, etc. Yet in fairness to Jeffrey, it should be pointed out again that he does not see his theory of preference as a "branch of English linguistics ... ," but as a possible way of interpreting *talk* reflecting practical concerns in the everyday world.[12]

However, almost at the same time Jeffrey is seen to claim that propositions are linguistic entities, or that they have a strong "affinity" to them. Yet, what does he mean by a "linguistic entity?" Can one conceive of a proposition other than in relation to some linguistic expression, within a specific system of language? Jeffrey's explanation of what he means by this term does not go very far in answering these questions. For example, he speaks of propositions being "named" by putting the word "that" or the expression "the proposition that" ahead of the corresponding declarative sentence. In doing this, he draws the distinction between proposition and sentence as being that of the difference between direct and indirect quotation. In Jeffrey's view, to refer to a sentence in the direct voice is to make reference to that particular sentence which some-

---

[11] Jeffrey, Richard C., *The Logic of Decision*, p. 60.
[12] ibid., pp. 63-64.

one utters, as for example, he said that "The sun will rise tomorrow." This is an illustration of the case where the "that" term is used to introduce the name of a sentence. On the other hand, where one speaks in the indirect voice, one is in the position of naming a proposition. For to say he said that the sun will rise tomorrow can be taken as saying the same thing as he said that tomorrow there will be daybreak. In this case, the "that" term is used to name a proposition, which is what the uttered sentence expresses, and it can also be seen that while the direct voice discriminates between sentences, the indirect voice does not differentiate between the same proposition which may be expressed by two different sentences. Finally, for purposes of his own analysis, Jeffrey stipulates that the logic of preference he will be developing will consider the objects of "belief" and "desire" as propositions, i.e., as expressions of sentences taken in the indirect voice. Thus, he says that propositions are seen to be the objects of both of the following sentences:[13]

> Columbus believed that the earth is round.

and

> From Columbus' point of view it was desirable that
> it should be the case that the earth is round.

In both of the above sentences, it is propositions as the objects of desire or belief which form the subject matter or object of the believing and desiring.

Jeffrey goes on to claim a distinct advantage to the procedure of holding propositions instead of sentences as being the objects of beliefs and desires. He claims that though the sentences, "It is not both raining and snowing," and "Either it is not raining or it is not snowing," both express the same proposition, it may well be that some agent may believe the first and not the second. For he might well assent to the first without knowing that it is "logically" equivalent to the second, and therefore withhold his assent. He notes that a similar point can be made about the agent's desiring what the first sentence expresses, and yet claim not to desire what the second sentence is saying. All in all, he argues that it is propositions which are the "appropriate" objects of belief, and *not* sentences. Thus, though one may have a willingness to assent to propositions, this does not mean that one may assent to the sentences which express the propositions.[14]

Jeffrey points out as well that though we commonly express propositions by means of sentences, one should not infer that one believes only what

---
[13] ibid., pp. 64-65.
[14] ibid., pp. 65-66.

is stated in a sentence, or that one cannot be said to believe in anything, unless a sentence is considered *as uttered*, or as somehow before the believer. Surely, Columbus believed that the world was round without knowing English, and one may say of an animal's behavior that it believes it is time for it to be fed. Yet, neither Columbus' belief requires the necessary utterance or presence of a sentence, nor does one accept that the animal has framed a proposition in its supposed belief that it will be fed. All one has here, according to Jeffrey, is our own imposition of a theory of deliberative action upon the phenomena we are observing, for the sake of interpretation. The practice of projecting a model on what is behaviorally manifested arises because ordinary talk finds it convenient and natural. Moreover, Jeffrey goes on to raise doubts concerning the reliability of questioning for the purpose of determining belief, since this is only a "rough and ready" means of specifying belief, which often the respondent is unwilling or unable to account for through his answering. Thus, Jeffrey concludes this section of his work by observing that though the criteria of belief and desire are behavioral, we must not lose sight of the fact that language is only *part of* that behavior. Consequently, one must have a broader picture of the kinds of things which qualify as the objects of what an agent believes, and that one's reduction of the objects of desires and beliefs to propositions is but one possible way of characterizing that belief.[15]

In a recent article, "Animal Interpretation," Jeffrey takes a stronger stand in his insistence that lower forms of life express preferential choice. In opposition to Davidson, he argues that insects, for example, can be seen as behaving in an "intentional" manner, with clearly evident manifestations of desires and beliefs. In Jeffrey's view, bees, in attempting to remove the remains of one of their own, exhibit a marked degree of "rationality," quite independently of whether it could be said of them that they can test, consider, reject, or accept hypotheses. He reasons that all biological creatures are animated by four fundamental prerequisites of survival: feeding, fighting, fleeing and procreating. Consequently, every situation requiring the recognition of desires, as well as the recognition of the satisfiers of desires, within the context of the four basic drives outlined above, manifests what can be termed *the preferring* of one course of action as opposed to another, whether this be attributable to humans or to non-humans. In this respect he claims that talk about preference is grounded in "prelinguistic soil."[16]

---

[15] ibid., pp. 60-70.
[16] Jeffrey, Richard C., "Animal Interpretation," (manuscriptcopy), pp. 6-7.

In contrast to Davidson, Jeffrey proceeds to argue that though we identify preferential states in terms of sentences expressing desires and beliefs, it is equally correct to claim that animals also prefer, in that they have expectations and wants, and that this can be maintained without committing the absurdity of saying that animals understand human speech. It is from the interpretation or reading of animal behavior that one infers that they desire to be fed, or that they want to drink. Hence, on the level of first-order preference one should include animal action, and in doing so accept what Davidson is unwilling to consider, namely, that at this basic level, rationality is attributable to animals precisely because they can be seen to exhibit degrees of belief. Here Jeffrey makes the very interesting distinction between human rationality and animal unrationality as meaning in Davidson's analysis the difference between *autonomy* and *automatism*. It is exactly on this crucial point that Jeffrey and Davidson differ widely.[17] For Davidson reserves the attribution of rationality only to creatures like man, who have the capacity to criticize and alter their beliefs, whereas lower forms of life are said to be motivated exclusively by mechanical and, therefore, inherently unreflecting reactions to stimuli.[18] Jeffrey, on the other hand, while recognizing the strength in Davidson's characterization of the basis of human rationality, presses the point that a significant portion of human action can also be seen to be automatic, though one would scarcely hasten to identify such action as irrational, e.g., reaching for an umbrella on a rainy day. Consequently, it in no way seems necessary that conscious awareness of *sententially expressed choice* is a fundamental prerequisite for the presence of preferential action. Herein lies the "prelinguistic soil" of preference.

Though Davidson's departure from Jeffrey will be given closer attention furtheron, a point needs to be made here relative to Jeffrey's argument against Davidson. For it seems that in "Animal Interpretation" Jeffrey has shifted ground from a discussion about the *un*rational to a discussion about the *ir*rational. Surely, Davidson would not claim that mechanical reaction to stimuli is *ir*rational action, but that it may be *un*rational action. The later is simply the manifestation of action without the immediate interjection of critical doubt. Surely, humans can be said to act in such a manner on many occasions in daily life. Yet for humans it is a mode of behavior which has been derived from some past reasoned activity, however simple its formation. This automatic mode of action is the result of having reasoned that such and such an action is produc-

---

[17] ibid., pp. 7-8.
[18] Davidson, Donald, *Inquiries Into Truth and Interpretation*, Clarendon Press, Oxford, 1984, pp. 164-165.

tive, and which through successful application has now become automatic in the sense that one is presently unaware of the original reasoning which led to its implementation. Such action is unrational to the extent that in its present manifestation it requires no reasoning. However, such action is not irrational, in the sense that there is absolutely no reasoning process connected to it. Davidson argues that animals cannot be said to act in such an unrational manner because they cannot reason about their choices, in that they lack the proper understanding of speech and, therefore, of the rational ordering of options. Nor are their automatic responses properly said to be irrational either, in the sense that this presupposes *the capacity* to choose between clearly expressed alternatives, though there is a failure to do so. Rather, Davidson will be found to argue that apart from a basically mechanical reaction to stimuli, one could not impose upon an interpretation of animal activity any kind of explanation involving human reasoning processes. To do so would constitute projecting an anthropocentrically oriented interpretation of behavior upon a lower form of life.[19]

A further ramification of Jeffrey's remarks up to this point illustrates how wide of the mark from Martin's approach he allows his views to proceed. For whereas Martin presents an analysis of belief in terms of an extensionalized pragmatics of testing and acceptance, where the objects of belief are taken as sentences considered in extension, Jeffrey would see this as presenting a much too limiting if not inaccurate perception of what "belief" involves. Surely, Jeffrey is saying in the above that one must be sensitive to the fact that the objects of beliefs are not limited to the particular characterization they receive when considered in terms of their linguistic extension. To this, one could argue that the approach embarked upon by Martin is meant to elucidate the nature of belief claims, and as such it is only meant to expose their most pervasive facets, which would not have been discernible by an intentional mode of analysis.

However, the point here should not be to defend Martin's approach but rather to assess Jeffrey's views on propositions, and the role they play as "objects," so-called, of beliefs and desires.

In scrutinizing Jeffrey's conception of proposition, there appears to be a conflict between his preliminary contention that propositions somehow have an "affinity" to linguistic entities, and what he says toward the end of his discussion on this subject, where propositions are seen as not needing to have

---

[19] ibid., pp. 155-156.

corresponding sentences expressing them, since it is propositions which form the objects of beliefs, and not sentences. Thus, one seems to have at the end of his account a notion of proposition as somehow distinct from a sentential expression, and also as something which may be without any linguistic affinities, as in the case of the animal's believing that it will be fed. Curiously, by his last remarks concerning the difference, so-called, between belief and assent, his conception of propositions begins to suffer woefully. For he says that assent can be given to propositions if one construes this as assent being given to the sentence which expresses the proposition. Hence, one seems to have here a meandering view of proposition. On the one hand, propositions are said to have an affinity to linguistic entities, then they are said to be independent of linguistic expression, and still later they are held to be "assented to" *via* the intercession of sentences.

In light of his concluding remarks concerning the broadness of the notion of proposition, how is one to interpret what he has to say where he claims that one desires that the proposition about a good woman's love toward him "holds," in the case where one contemplates the desire of a good woman's love?[20] What is this "holding" which the proposition must do in the attitude of desire? Does it mean that the sentence which expresses this proposition must be true or verified to be the case factually? Surely, it is not the proposition which is said to "hold" here, but the sentence, which alone can be said to be true or false. If in Jeffrey's conception a proposition is in itself independent from any locutionary attitude, i.e., assertion, questioning, commanding, etc., then what sense does it make to say that one can desire that a proposition "hold"? The whole basis of Jeffrey's logic of preference, as a unified theory which treats desires and subjective probabilities on the same level as propositions, is now seen to be suspect, because of his failure to clearly explain the notion of a proposition and of its truth.

The status of the notion of proposition as employed by Jeffrey becomes even more problematic where it is said to be relevant to actions. In Chapter 5. he claims that actions can be expressed in terms of declarative sentences, e.g.,

>We have red wine with dinner.
>The agent takes the plane.
>The agent disarms.

In these cases one has an agent, be it an individual or a state (as the last sen-

---

[20] Jeffrey, Richard C., *The Logic of Decision*, p.59.

tence suggests), who attempts to perform the act indicated in each sentence. He stresses the point of the "attempting" or the "trying" to perform the actions expressed by these sentences, for the "trying" mode, so to speak, preserves the general probabilistic attitude which underlies his logic of preference. However, he goes on to say that if the sentence accurately characterizes the act, then the act can be "conveniently" identified with the proposition which the sentence expresses. In a crucial passage, he states: "... An act is then a proposition which is within the agent's power to make true if he pleases, and the necessary proposition would correspond to not acting; to letting what will be, be."[21] It is this very identification of acts with propositions, which he goes on to characterize as a "realistic" identification, which seems so tenuous. His rationale for introducing this discussion is to provide a means of having an active and passive notion of preference. For where an agent thinks that he can perform the act required by any two propositions, and proceeds to rank doing one act as opposed to the other, this is indicative of an active sense of preference. However, where there is no ranking of actions temporally in terms of what will be done first, there one has the preference of taking no action, which is Jeffrey's idea of passive preference.

The issue of the tenability of the claim that propositions can be properly said to "hold" seems to have been missed even by Donald Davidson, in his review of Jeffrey's work in the essay "Belief and the Basis of Meaning." Davidson endorses the former's attempt at clearing up Ramsey's "rather murky ontology" when presenting his theory of preference. Alluding to the 1965 edition of *The Logic of Decision*, Davidson comments upon Jeffrey's attempt to present a unified theory of preference. In an interesting passage Davidson says: "... Preferences between propositions *holding true* (my italics) then becomes the evidential base, so that the revised theory allows us to talk of degrees of belief in *the truth of propositions* (my italics), and the relative strength of desires that propositions *be true* (my italics). ..." For Davidson sees no great need for explaining just how truth is to be assigned to propositions, nor does it seem to him a terribly difficult thing to say that propositions can be taken as "holding true." Remarkably, in the same essay Davidson is found to be concerned over the vagueness of the properties of propositions in relation to those of numbers, and he therefore questions the sense of saying that propositions are the meanings of sentences or "objects" of belief. Such concern seems out of place with his willingness in the quoted passage to follow Jeffrey in the latter's assump-

---

[21] ibid., pp. 83-84.

tion that propositions can be said to hold true, as if no question existed as to the definiteness and clarity of the concept of proposition. Moreover, almost in the same breath Davidson notes how all of what Jeffrey has to say about preferences with regard to propositions could also be said about preferences between sentences. Davidson finds no difficulty in making the transition from propositions to sentences, though he is seen to be troubled by the vagueness of "propositions" as a philosophical notion.[22] In all of this one wonders whether on a conceptual basis Jeffrey has succeeded in clearing the murky waters Ramsey has left behind.

Another real problem here is with Jeffrey's allusion to the way in which the above sentences are said to "describe" an action, and that the agent acts upon the proposition which is expressed by the sentence, so that the proposition is made true in virtue of some act. Again what sense does this make? How can an act make true a proposition? Is it by the utterance of the sentence "We have red wine with dinner" that one is describing an action? Yet, in this relatively simple sentence there is described not one action, but a whole series of them, e.g., the agent bringing in the red wine, the pouring of the wine, the savoring of it, the eating of the dinner, etc. Thus, to what *single* proposition is the action directed, if there are several discernible actions described? Furthermore, in the previous chapter propositions were said to be in some cases independent of sentential expressions, as in the case of the dog's believing *that* it will be fed. In cases such as the last, one questions whether any *one* action is being described by such propositions. Surely, to claim that a description has taken place is to suggest at least some definite parameter where the description begins and where it ends. Yet it is difficult to see how this could take place without some strong affinity with a sentential form. In essence, one may well ask Jeffrey how a description can take place without symbolic expression of a coherent and interpersonal nature. To claim that a description has taken place, or is taking place, implies that ambiguity has given way to clarity, and that some indeterminate has become definite. This, at the very least, seems to be the heart of saying that something has been described. Yet to divorce propositions from sentences, as Jeffrey is found to do, and to claim that such propositions can describe actions, and that they can be held to be affected by actions, seems to be virtually beyond one's grasp.

One could allude once more to Davidson on this score, with a somewhat better result stemming from his 1975 essay entitled, "Thought and Talk." In a

---

[22] Davidson, Donald., *Inquiries Into Truth and Interpretation*, pp. 147-149.

concluding passage, Davidson voices an opinion concerning the correct view of what is to be construed as an "object" of belief. His remarks have the same ring as those seen in Alan White's view on the central role of language in the expression of meaning. White's views have been considered in greater detail in connection with Rescher's attempt to evolve a highly mathematical rendition of the logic of preference. Here Davidson's observations are found to clearly echo White's feeling that one simply cannot separate the act of uttering or writing from the sense or meaning of what is expressed. This point is especially germane to the view which is so central to Jeffrey's thesis, namely that there is a legitimate claim in saying that a proposition as the sense or meaning of a sentence can be divorced from the sentence itself, and that a lower form of life can apprehend meaning independently of linguistic expression as we know it. Davidson is found to say as follows:[23]

> We have the idea of belief only from the role of belief in the interpretation of language, for as a private attitude it is not intelligible except as an adjustment to the public norm provided by language. It follows that a creature must be a member of a speech community if it is to have the concept of belief. And given the dependence of other attitudes on belief, we can say more generally that only a creature that can interpret speech can have the concept of a thought.
>
> Can a creature have a belief if it does not have the concept of belief? It seems to me it cannot, and for this reason. Someone cannot have a belief unless he understands the possibility of being mistaken, and this requires grasping the contrast between truth and error - true belief and false belief. But this contrast, I have argued, can emerge only in the context of interpretation, which alone forces us to the idea of an objective, public truth. ...."

What is not available to Jeffrey is the kind of argument proposed by Peter Geach where he says that a proposition is "... A form of words in which something is propounded, put forward for consideration, it is surely clear that what is being put forward neither is *ipso facto* asserted nor gets altered in content by being asserted. ..." Geach exemplifies what he means here by alluding to the expression in formal logic, "if p, then q," which makes absolutely no assertion about *what* 'p' or 'q' stand for. Here one has the case of the assertion of a hypothetical proposition within the language of formal logic, without being com-

---
[23] ibid., p. 170.

mitted to the claim that it is either true or false, since the assertion itself is truth-functional.[24] However, for Jeffrey, what is desired is seen to be a proposition one wishes to "have" hold. It is this "having" which renders the object of desire, although a proposition, something which must or must not (in an exclusive sense) obtain either in the agent's present, or at some future time. For this reason, one simply cannot treat the sense of proposition expounded by Jeffrey in his explanation of the notion of desire in the same vein as Geach's sense of proposition, which though put forward is still neither factually true nor false.

The foregoing provides for some additional points of criticism as to Jeffrey's method of operation. Clearly, Jeffrey bases preference on desirability, and conversely. The desirable is to be somehow characterized numerically within a context of publicly evaluate desire by some agent. Consequently, preference is defined as the desirable, relative to the probability of that which is desired coming true. However, what is not fully explained by Jeffrey, as was also seen to be the case with the attempt by Nicholas Rescher, is the role of the agent's needs in determining preference. This is to say that Jeffrey presents a largely "derivative" sense of preference, meaning that he does not see preference-ordering as a distinct (i.e., independent) pragmatic activity apart from the desirable, but rather it is directly dependent upon the latter, as well as upon the probability of the occurrence of desired states of affairs. Jeffrey assumes that numerical attribution relative to some desired proposition suffices to explicate effectively the notion of preference-ranking. Yet this kind of procedure does not resolve the issue of whether in fact all cases of preferring are animated by the core notion of desirability as Jeffrey supposes. Noticeably absent at the outset of his work is any allusion to the kind of instructive differentiation one encounters in von Wright's *The Logic of Preference*, with respect to intrinsic and extrinsic preference. Apparently, Jeffrey is operating with a concept of preference which is extrinsic, i.e., a conception of preference where observation and risk is always possible. The thrust of Jeffrey's effort is to take the numerical attribution of the desirable as given, with the tacit understanding that it is also a 'testable' factor.

Much remains unexplained in claiming that the concept of preference is derived from the notion of the desirable, and that preference in this sense is also open to an extensional mode of analysis At this point it is worth mentioning that "to desire," as an expression in *ordinary speech*, carries along with it a strong connotation of immediacy, in that the object desired is often seen as

---

[24] Geach, Peter., "Assertion," in *Readings in the Philosophy of Language*, Jay F. Rosenberg and Charles Travis, eds., Prentice Hall, 1971, pp. 252-253.

what should or must be possessed by the agent, regardless of any impediment standing in the way of his acquisition of it. This, at least, is the intrinsic sense of desire encountered in von Wright's investigations. Somewhat similar observations were seen to be made by Brentano, where the latter observed that the preferable is that state of the desirable whose intensity is intuitively recognized as the highest or most excellent condition to achieve. Moreover, as von Wright correctly points out, it is only in the intrinsic sense of preference that one is justified in speaking of the preferred as reflecting the desirable. When extrinsic considerations are introduced, and the possibility of *risk* enters the picture, then preference should be considered in terms of the agent's needs and projected goals rather than desires. The latter in ordinary parlance usually assumes the stipulation of things in the agent's environment being non-consequential. On the other hand, it is precisely because of the highly unpredictable character of preference in its extrinsic sense that von Wright had elected to develop a logic of preference in the more "stable" intrinsic sense, i.e., of preference as it reflects what is determined to be the desirable *in and of itself*, all things being equal. The crucial point which is emerging here is that on the one hand Jeffrey is concerned with preference in an extrinsic sense, a sense which lends itself to experimental determination and quantitative characterization, and yet on the other hand he insists on interpreting preference in the mode of the *desirable*, which ordinarily is separate from considerations involving risk and/or one's probable success in the acquisition of an end. In summary, it seems that there is something wrong in the way in which Jeffrey is attempting to "clarify" ordinary talk about preferences, while also developing a logic which takes risk into account, as well as the probable in any course of action.

One could argue, however, that probability theory is essentially context-free, meaning that the statistical correspondences which it deals with are seen within a mathematically coherent background, and that one should not view the assignments of values to states of affairs as indicative of a description of subjective states of mind. Here one may well recall Bruno De Finetti's incisive observation that "... it is essential to point out that probability theory is not an attempt to describe *actual* behavior; its subject is *coherent* behavior, (italics added), and the fact that people are only more or less coherent is inessential." To this Henry Kyburg makes an additional observation that no amount of the testing of people will lead to evidence which will confirm the truth of a subjectivist theory of probability. For such inquiries, Kyburg continues, only illustrate how people are prone to act incoherently on occasion, which is not in itself evidence for saying that the probability theory under consideration is or is not adequate. In summary, it can be said that subjectivist probability theory can be taken as an example of how individuals *should* act, though it is not indicative of how they *do* act.[25]

However, whereas the cleavage between the dictates of probability theory and particular human activities is defensible upon purely conceptual grounds, the matter is somewhat different when it comes to formalizing preferences, which are uniquely human acts. In the case of the latter, there is the conscious effort to involve on a conceptual level notions such as desire or the willingness of an individual to act in a particular manner, as well as ideas reflecting a person's needs and relevant goals. Clearly, terms such as "desire," "willingness," "needs," etc. have a uniquely person-relevant connotation, and it is very much of an open issue whether one can express in terms of probability theory the *signification* these terms have in ordinary discourse. One must again be mindful of the fact that one of the innovative aspects of Jeffrey's work in *The Logic of Decision* is to face the issue of evolving a logic of preference which is in touch with the way preferences are spoken of in "ordinary talk". Moreover, his discussion of propositions in the preamble to his logic of preference is an additional thrust in the direction of focusing upon the role of language use in guiding the emergence of his theory. With a few exceptions, Jeffrey's approach is unique in that it is sensitive to the need that there be a uniform ontology underlying a logic of preference, which he feels he has provided by attributing to *propositions* both subjective desirability and subjective probability. Jeffrey's commitment to a logic of preference, which is descriptive of the way in which ordinary discourse expresses the preferential mode, is thus essential in seeing the intention behind his theory in *The Logic of Decision*. In this connection one could argue that unlike most efforts discussed thus far, Jeffrey cannot be interpreted as trying to show how one "should" rationally prefer, rather he is more concerned with illustrating how our talk about preference can be formally characterized.

In the majority of cases encountered thus far, one sees how an attempt is made to describe what preference *should be* within a formal system, where the system itself dictates the criteria for the logic of preference. Specifically, there is here the problem of how to justify saying that coherence among preferences within a logic is or should reflect in some ideal or puristic sense the coherence of preference, which is presumed to be operating in everyday circumstances. Is systemic coherence an applicable criterion for the proper preference-ranking

---

[25] Kyberg, Henry E. jr., *Studies in Subjective Probability*, John Wiley & Sons, Inc., and Smokler, Howard E. New York, 1964, p. 111 and p. 6.

of "ordinary" states of affairs? What is the function, if any, of a criterion in such cases? If the purpose of such logics is to illustrate what coherent preference *ought* to be like, then what is the relevance of this kind of coherence to the broader and often unwieldly context of human actions, where preferring may be manifested? Should one disown the "ordinary" sense of preferring, or is any logic of preference proffered thus far, however abstract, all that one really means when expressing preferences in ordinary discourse?

Answers are not easy to come by here. There neither appears to be nor does it seem necessary that there be a clear point of tangence between coherence in one context, and coherence in the other. In fact, one may doubt whether it makes any sense to speak of the coherence of preference in an "ordinary context," so-called, since knowledge of all the variables necessary so as to determine coherence would come close to being impossible. This would be evident especially where one is taking preference extensionally, where the role of need must be introduced as forming the context of preference. Surely, the complete enumeration of needs becomes a virtually impossible task. It would require the omniscience of being able to foretell the future course of the long range consequences of one's choice. Simply, the point here is that "coherence" as the degree of integration a statement has in relation to an entire body of internally ordered claims seems difficult to apply when it comes to considering how preferences are ordered within the context of everyday circumstances.

If any one conclusion can be drawn from the above observations then it is that there appears to be no single logic of preference per se, but rather there are many such logics, depending on one's conception of preference and the conditions under which it is manifested. Consequently, one may be putting too much importance upon a structured sense of coherence, were one insists that probability analysis can demonstrate how one should act when it comes to preference-ranking.

The issue, of course, can be forced in that one can question the very sense of saying that within the highly structured context Jeffrey's logic provides, one can have a means for intelligently expressing desire, or even preference. Though a similar point has been raised on other occasions in the past, it is again worth mentioning. Surely, it is very much an open question whether a rigorously defined mathematical context can provide an effective means of clarifying the sense of preference employed in everyday circumstances. It would appear that the most which could be said of what preference in a struc-

---

[26] Jeffrey, Richard C., *The Logic of Decision*, p. 80.

tured sense means is the numerically highest (or as the case may be lowest) value. Yet, again if one were to interpret this numerical representation as the "preferred" state of affairs, it can at best only roughly approximate all that the preferred could mean in an ordinary language context.

An additional point is worth pursuing prior to continuing with the exposition of Jeffrey. This is the observation that it apparently makes more sense to concern oneself with consistency *among* preferences, rather than with their coherence. This seems more in keeping with the emerging realization that one seems unable to arrive at a single, all-inclusive, logic of preference, and that perhaps the most that could be hoped for is a formalization of preference-relations which takes into account only particular situations in which preferences are found to occur.

Apart from the difficulties one can point to in attempting to scrutinize the role of desire within a formally defined setting, it is well to keep in view that Jeffrey interprets the desirable as some sententially expressed proposition, and as such it is held to be either true or false. This also means that one simply accepts that in the context of Jeffrey's mode of inquiry, the desirability of states of affairs, expressed by means of a numerical designation, is a given from which certain formal inferences follow. In this connection, one should distinguish between numerical designation indicating desirability, from numerical designation indicating the probability of the occurrence of these states of affairs. The latter constitutes a totally different frame of reference. This, in a manner of speaking, is the minimal base from which Jeffrey launches his investigation. Matters pertaining to private states of consciousness, as these relate to the desirable as a specific state of consciousness, simply are not seen by him as falling within the purview of his inquiry.

In line with the above, one finds Jeffrey presenting three basic axioms of the probability calculus which are fundamental to his discussion, together with a "desirability axiom," which results in an elementary desirability calculus. He emphasizes the point that the term 'calculus' here is meant to signify in both instances a method for calculation. These axioms are:[26]

(a) *prob is nonnegative*: $prob\ X \geq 0$
(b) *prob is normalized*: $prob\ T = 1$
(c) *prob is additive*: if $XY = F$, then $prob\ (X \vee Y) = prob\ X + prob\ Y$.

---

[26] Jeffrey, Richard C., *The Logic of Decision*, p. 80.

The first axiom asserts that probability cannot be negative. The second claims that the probability of all necessary propositions is 1, and finally the third axiom asserts that the probability of two incompatible propositions is additive.

To this Jeffrey adds the desirability axiom which simply says that:

"if $prob\ XY = 0$, and $prob\ (X \vee Y) \neq 0$, then $des\ (X \vee Y) = \dfrac{prob\ X\ des\ X + prob\ Y\ des\ Y}{des\ X + prob\ Y}$."

These four axioms are said by Jeffrey to yield precisely the same result as the rule stated earlier regarding the computation of desirabilities as the weighted average of the desirabilities of the cases in which they are true.

Jeffrey turns next to a way of expressing the desirability of necessary propositions, or T propositions, as noted above. In line with his previous analysis, he expresses the desirability of T as the weighted average of the product of X and of its desirability, in sum with the product of the desirability of $\bar{X}$, and the probability of $\bar{X}$, i.e., "$des\ T = prob\ X\ des\ X + prob\ \bar{X}\ des\ \bar{X}$". Since T must be true in all cases, it will be ranked along with X, where the probability of X is 1. However, T will be ranked with $\bar{X}$ where the probability of X is 0. In the case where the probability of X is neither 1 nor 0, then T will be ranked between X and $\bar{X}$.[27]

Jeffrey is now in the position of presenting a means of "appropriately" distinguishing between "good," "bad," and "indifferent," within his calculus of preference. Given proposition A, then one could say that A is good if $des\ A > des\ T$, and A would be bad if $des\ A < des\ T$, and finally it can be said that A is indifferent if $des\ A = des\ T$.[28]

Alternatively, one could interpret $des\ A > des\ B$ as meaning that A is preferred as a "news item" to B. This suggests that some agent would welcome the truth of A over the truth of B. Consequently, a proposition T, which expresses an impossibility, would not occur within a preference-ranking because it cannot make news. In this case, to rank T above A means that no news is better than bad news, i.e., A. Where A is ranked above T, one has the expression of good news, i.e., A, is better than no news, i.e., T, which is now bad news. Finally, where A and T are ranked evenly, one has the expression of the agent being indifferent to any news.

---

[27] ibid., pp. 81-82.
[28] ibid., pp. 83-84.

140   CHAPTER 4

In the final section of the fifth chapter, Jeffrey notes an unusual result of his logic of preference thus far. He observes that a proposition which is said to be true in more than one case is a *gamble* between those cases in which it is said to be true. Consequently, one sees how the probability of such a proposition X can be expressed solely in terms of its desirability. For example, one can compute the probability of X from the expression of the desirability of T, i.e.:

des T = (*prob* X)(*des* X) + *des* X̃ - (*prob* X)(*des* X̃), where *prob* X̃ =1 - *prob* X. Thus where *des* X ≠ *des* X̃, and T expresses (X v X̃), one has:

$$prob\ X = \frac{des\ T - des\ \tilde{X}}{des\ X - des\ \tilde{X}}$$

In cases such as these, one readily sees how within the parameters of Jeffrey's work it is desirabilities which determine probabilities. Keeping the proper perspective on this result, however, requires that one carefully note how X must first be found to be a proposition which may be true in *more than one case*, i.e., a gamble, *before* it can be said that in such a case probability is determined by desirability.[29]

Jeffrey turns next to the formal presentation of equivalence between preference-ranking. To this purpose he makes use of Ethan Bolker's Equivalence Theorems as a means of giving a "partial" description of what counts as equivalence between two pairs of probability and desirability assignments, i.e., "*prob, des*" and "*PROB,DES*", respectively.

The situation to be described is expressed by the two following equations:

"(6-1) PROB X = (*prob* X)(c *des* X + d)," and

"(6-2) DES X = $\frac{a\ des\ X + b}{c\ des\ X + d}$."

In the above, the constants *a* and *c* "need not" be positive numbers.[30]

Prior to the application of the *equivalence theorem*, Jeffrey defines what he means by the *existence condition* which the pairs *prob, des* must meet relative to a given-preference ranking of propositions. The purpose served by the existence condition requirement is to have a criterion by which to argue that if say pair *prob, des* meets the existence condition, and *PROB,DES* also meets this condition, then *prob, des* and *PROB,DES* are equivalent. This condition requires that relative to a preference-ranking the pair must meet the probability and de-

---
[29] ibid., pp. 85-86.
[30] ibid., p. 96.

sirability axioms noted above, and that in addition the second member of the pair, i.e., the desirability component, *des X*, must "mirror" the preference-ranking: "in the sense that *des X* is *at least* as great as *des Y*, whenever *X* is ranked as high as *Y*. ..."[31] Therefore, in rough summary one could say that the existence condition serves to insure a basic parity in preference-ranking between any two distinct pairs. It should also be observed here that this is *not* held as an identity relation between preference-ranking.

With the above as preparation, one can turn to Jeffrey's use of Bolker's theorem.

The theorem states quite simply that if 6.1 through 6.2 above are satisfied by a given pair *prob, des*, and this same pair also meets the following conditions:[32]

"(6.3) (a) $ad - bc$ is positive
(b) For each X in the preference-ranking, $c\ des\ X + d$ is positive
(c) $c\ des\ T + d + 1$. ..."

and *des, prob* also satisfies the existence condition then the following five requirements are met, and therefore *PROB,DES* also meets the existence condition. These requirements are:[33]

"(i) The values of *PROB* are never negative;
(ii) $PROB\ T = 1$.
(iii) $PROB\ (X \lor Y) = PROB\ X + PROB\ Y$ if $XY = F$.
(iv) if $PROB\ XY = 0$ but $PROB(X \lor Y) \neq 0$, then
$$DES\ (X \lor Y) = \frac{PROB\ X\ DES\ X + PROB\ Y\ DES\ Y}{PROB\ X + PROB\ Y}$$
(v) $DES\ X \leq DES\ Y$ if and only if $des\ X \leq des\ Y$ ...."

The verification of (i) can be had in three steps as follows:

1. $c\ des\ X + d$ is positive for every X in the preference-ranking: (6-3)b.
2. $PROB\ X = (prob\ X)(c\ des\ X + d)$: (6-1)
3. *prob X* cannot be negative: (Jeffrey s first axiom (a)).

Requirement (ii) is proven as follows:

1. X is set at T, i.e., $X = T$, (by assumption).
2. $PROB\ T = (prob\ T)(c\ des\ T + d)$, (6-1)
3. $c\ des\ T = d = 1$, (6-3)c.
4. $prob\ T = 1$, (Jeffrey's axiom (b)).

---
[31] ibid., p. 96.
[32] ibid., p. 97.
[33] ibid., pp. 97-99.

Requirement (iii) involves the consideration of the following two cases:

In the first $prob(X \text{ v } Y)$ is taken as equaling 0. Consequently,

    1. $prob\ X = prob\ Y = 0$, (Jeffrey's third axiom (c)).
    2. $0 = 0 + 0$, (application of step 1. to (iii) above) (This proof seems somewhat more cogent than what appears in Jeffrey's text at this point.)

In the second case requirement (iii) is demonstrated on the basis of the assumption that $prob\ (X \text{ v } Y) \neq 0$.

    1. $PROB\ (X \text{ v } Y) = prob\ (X \text{ v } Y)\ [c\ \dfrac{prob\ X\ des\ X + prob\ Y\ des\ Y}{prob\ (X \text{ v } Y)} + d]$
(application of the axiom of desirability to $prob\ (X \text{ v } Y) = 0$).
    2. $= c\ prob\ X\ des\ X + c\ probY\ des\ Y + d\ prob\ X + d\ prob\ Y$,
(the natural reduction of step 1.).
    3. $= (prob\ X)(c\ des\ X + d) + (prob\ Y)(c\ des\ Y + d)$,
(application of the desirability axiom to step 2. above).
    4. $= PROB\ X + PROB\ Y$,
(application of formula (6-1) to line 3.).

The verification of the fourth requirement proceeds on the assumption that $PROB\ XY = 0$, and that $prob(X \text{ v } Y) \neq 0$. Consequently,

    1. $(prob\ XY)(c\ des\ XY + d) = 0$, (applying equation (6-1) to $PROB\ XY = 0$).
    2. $prob\ (X \text{ v } Y)(c\ des\ (X \text{ v } Y) + d) \neq 0$, (applying (6-1) to $prob\ (X \text{ v } Y)$.
    3. $DES\ (X \text{ v } Y) = \dfrac{a\ des\ (X \text{ v } Y) + b}{d\ des\ (X \text{ v } Y) + d}$,
(application of (6-2) to $PROB\ XY = 0$).
    4. In line 3. Jeffrey employs the axiom of desirability to both the numerator and to the denominator, resulting in:
    5. $DES\ (X \text{ v } Y) = \dfrac{(prob\ X)(a\ des\ X + b) + (prob\ Y)\ (a\ des\ Y + b)}{(prob\ X)(c\ des\ X + d) + (prob\ Y)(c\ des\ Y + d)}$.
    6. $DES\ (X \text{ v } Y) = \dfrac{PROB\ X\ DES\ X + PROB\ Y\ DES\ Y}{PROB\ X + PROB\ Y}$,
(application of (6-2) and (6-1) on line 6.).

The fifth requirement is verified by observing how by multiplying an inequality by the same positive number, the resulting inequality is equivalent to the original inequality. Thus $DES\ X \leq DES\ Y$, if one has the application of (6-2), as follows:[34]

1. $\dfrac{a\ des\ X + b}{c\ des\ X + d} \leq \dfrac{a\ des\ Y + b}{c\ des\ Y + d}$

2. $ac\ (des\ X)(des\ Y) + ad\ (des\ X) + bc\ (des\ Y) + bd$
   $ac\ (des\ X)(des\ Y) + bc\ (des\ X) + ad\ (des\ Y) + bd$:
   (multiplying by (6-3)(b) on line 1.).

3. $des\ X \leq des\ Y$:
   (applying (6-3)(a) to line 2., and multiplying by 1/ad-bc)).

4. $DES\ X \leq DES\ Y$:
   (4. is equal to the original inequality expressed in 1.).

Jeffrey's presentation thus far does not account for the possibility that though within the framework of Bolker's Theorem $DES\ A = des\ A$, and $DES\ B = des\ B$, that des and DES agree "as to the desirabilities they assign to *all* propositions" (my italics). So as to tighten this looseness, Jeffrey proposes the introduction of an additional equivalence, $DES\ C = des\ C$, where proposition C is ranked neither with A nor with B.

In light of the above, Jeffrey proceeds to point out that by supposing the ensuing equivalences:[35]

$$(6\text{-}4)\ DES\ T = des\ T = 0 \qquad DES\ G = des\ G = 1$$

one can perform a series of transformations whereby setting X as equal to T in (6-1) one has:

$$1 = (1)(c.0 + d)$$

which yields:

$$d = 1.$$

The same setting of X as equal to T results in having (6-2) result in:

$$b = 0.$$

In the last step Jeffrey sets X to equal G, which with (6-2) obtains to:

$$a = c + 1.$$

This equivalence makes possible a reinterpretation of equations (6-3)(b),

For (6-3)(b) one now has (6-5): $c\ des\ X > -1$.

For (6-2) one now has (6-6): $DES\ X = \dfrac{(c+1)\ des\ X}{c\ des\ X + 1}$.

For (6-1) one now has (6-7): $PROB\ X = (prob\ X)(c\ des\ X + 1)$.

---

[34] ibid., pp. 98-99.
[35] ibid., pp. 99-100.

Also, assuming again that $X = G$, one has equation (6-8):
$$c + 1 = a \frac{PROB\ G}{prob\ G}.$$

By forcing c to zero in equation (6-5), (6-6) and (6-7) take on the forms of $DES\ X = des\ X$, and $PROB\ X = prob\ X$, respectively. Where c is bounded neither above nor below, equation (6-4) implies that the forementioned equivalences hold.

Jeffrey next entertains the possibility where *des* is not bounded either above or below or both, and c therefore need not be zero. In this case the issue is to determine what the values of c would be in relation to (6-5), where *des* has a certain range of values.

His method of answering this question is to introduce within a context of uniformly ranked propositions the ideas of *supermum* and *infimum* numbers. Respectively, the first as the number symbolized by s represents the least upper bound of the values of *des*. The second, symbolized by i, is the number of the greatest lower bound of the values of des. Where s and i may be infinite, ∞, or $-\infty$, four distinct possibilities emerge.[36] Namely,

1. *des* is unbound above and below, $s = \infty$ and $i = -\infty$,
2. *des* is unbound above but bound below, $s = \infty$, $i \neq -\infty$,
3. *des* is bound above, but unbounded below, $s \neq \infty$, $i = \infty$,
4. *des* is bound above and below, $s \neq \infty$, $i \neq \infty$.

Where X is a good proposition, then *des* X is positive, and (6-5) is written as:
$$c > -\frac{1}{des\ X}.$$
This inequality must hold whatever height proposition X is ranked, hence
$$c \geq -\frac{1}{s}.$$
Where X is a bad proposition, then des is negative, and (6-5) is written as:
$$c < -\frac{1}{des\ X}.$$
The above inequality must hold however low proposition X is in the ranking, hence
$$c \leq -\frac{1}{i}.$$

---

[36] ibid., pp. 101-102.

Combining the two inequalities results in the following equation (6-10):
$$-\frac{1}{s} \leq c \leq -\frac{1}{i}$$
Having set *des G* = 1, it is seen that *s* must be greater than 1, from which follows
(6-11) $\qquad c \geq -1$
for all preference-rankings conforming to the assumption of (6-4).

In the section on "Probability Quantization," Jeffrey considers the question of the degree of change in probability for proposition *X* by a transformation such as (6-7) above. The solution here depends on the value one assigns to *c*, as allowed by (6-10). Thus, *c* at its maximum value of
$$-\frac{1}{i}$$
results in the probability of *X* being expressed as:[37]
$$(prob\ X)\ [1 - \frac{des\ X}{i}].$$
Where *c* is at the minimum, the probability of *X* is expressed as:
$$(prob\ X)\ [1 - \frac{des\ X}{s}].$$
The difference between the minimum and maximum value assignments for *c* are seen in transformation (6-13):
$$prob\ X\ des\ X\ [\frac{1}{s} - \frac{1}{i}]$$
The issue now turns to expressing the degree of variation in probability as given in the transformation above. At this point Jeffrey introduces additional definitions for expressing the quantity *prob X des X*, which is symbolized by '*int X*' for the integral value of *X*. Jeffrey's new definitions are:

(6-14)   (a) *int X* = *prob X des X*,
        (b) *INT X* = *PROB X DES X*.

Unfortunately, Jeffrey confounds the next step in his proof, and what should appear in the order of applying definition (6-14) to the transformations of (6-1) and (6-2) respectively is reversed, so that (6-15)(b) below should be (6-15)(a), and conversely. In Jeffrey's text the transformations are given as follows:

(6-15)   (a) *INTX* = *a int X* + *b prob X*,
        (b) *PROB X* = *c int X* + *d prob X*.

---

[37] ibid., pp. 106-107.

Clearly, (6-15)(b) is derived from (6-1), and (6-15)(a) is derived from (6-2), once it is realized that $a = c + 1$, as in the derivation from (6-4) above.

Considering the Desirability Axiom (5-2) Jeffrey determines that the absolute value of variation for (6-13) must be less than 1. Consequently, he arrives at:

$$(6\text{-}18) \quad int\ X < \frac{1}{(1 \setminus 1s) - (1 \setminus 1i)}.$$

At this point, it is refreshing to reflect upon Jeffrey's intended purpose for the present chapter. As he claims in the introduction, Jeffrey wants to show how the preference-ranking of propositions determines both: (1) the utility function to a fractional extent, with a positive determinant, and (2) the probability function up to "within a certain quantization." In the case of (1), one sees in the rewriting of (6-5) above how he has been able to illustrate the fractional character of the utility function 1 as

$$c < -\frac{1}{des\ X}.$$

On the other hand, that preference-ranking can determine $des\ X$ the probability function up to a certain extent of quantization is illustrated by (6-18) above, where the integral number representing the quantity $prob\ X\ des\ X$ is taken as less than the difference between the minimum and maximum values of c. The latter constituting what Jeffrey means by the limiting or circumscribed "quantization," i.e. the minimum and maximum ranges of c.[38]

Jeffrey's work in the chapter entitled "From Preference toProbability" will be treated in more general terms than was the case with his comments in the chapter preceding the present one. Basically, it should be apparent that Jeffrey's logic of preference is well-established in the chapters reviewed thus far, and consequently much that follows by way of exposition of *The Logic of Decision* will be a summary of the major point she makes.

Jeffrey's work in the seventh chapter illustrate the ingenius integration of his mathematical analysis. He demonstrates with no uncertainty how one is "able to deduce features of the agent's probability assignment *from* his preference-ranking." What he is attempting to illustrate is what the probability assignments would be like given four distinct ranking conditions. These conditions are : (1) the *existence* condition, (2) the *closure* condition, (3) the G condition, and (4) the *splitting* condition.[39]

---

[38] ibid., pp. xii-xiii.
[39] ibid., pp. 116-117.

In the first case he states that there is a probability assignment for the given pair *prob, des* satisfying (5-1) and (5-2), with *des* mirroring the preference-ranking, such that given propositions A and B, in the ranking, *des* A is greater than, less than, or equal to *des* B, according to the ranking of A in relation to B.

In the case of (2) Jeffrey first defines what he means by a *probability field*. The latter is simply a collection of propositions which contains the denial of any of the contained propositions, as well as the conjunction and disjunction of any pair of propositions contained. He sets forth the closure condition as where the proposition in the agent's preference-ranking form a probability field, from which the impossible propositions have been removed.

In (3) one simple has the case where the preference-ranking contains a good proposition, G, for which its denial is bad. This condition says that given a preference-ranking containing a G proposition, the ranking for that proposition must be:[40]

$$G$$
$$T$$
$$G$$

since G is preferred to T, and the denial of G must be below T.

The final condition, which is termed the *Splitting Condition*, states that where A is in the preference-ranking, and neither A nor its denial are ranked with T, then there will be additional propositions, $A_1, A_2, \ldots A_n$, within this same ranking which will satisfy the conditions which follow:

"(1) $A_1 A_2 = F$;
(2) $A_1 v A_2 = A$;
(3) A_1 and A_2 are ranked together;
(4) A_1 and A_2 are ranked together."

The implication of this final condition is that any good or bad proposition having a positive probability can be split into two equiprobable propositions, and so on. Comparisons can then be made of two good propositions ranked together, as well as two bad propositions so ranked, in terms of the ratio:

$$\frac{prob\ A}{prob\ B}.$$

As stipulated in the presentation of the splitting condition, the procedure of splitting propositions cannot be directly applied where propositions are ranked with T. However, it can be applied indirectly to such propositions

---

[40] ibid., pp. 117-118.

to determine their numerical probabilities. This aspect of Jeffrey's analysis is interesting since it illustrates how "indifferent" propositions can be handled within his theory of preference. Briefly, what results from his discussion is that an indifferent proposition's probability is expressed as the disjunction of a specific number of divisions of G in disjunction with an equal number of divisions of the denial of G, or $\bar{G}$. This is, of course, a procedure which takes into account indifferent propositions exhibiting only the disjunctive form.

So as to accommodate the determination of the probability of indifferent propositions whose form are not disjunctive, Jeffrey devises an ingenious proof for a procedure which allows one to determine the probability of such propositions by comparing them to indifferent propositions having the disjunctive form. His first move is to present the notion of a "null" proposition. This kind of proposition is defined as:[41]

> "(7.2) A proposition A in the preference-ranking is null if and only if there is a proposition B in the preference-ranking for which we have
> (a) $AB = F$
> (b) $A \vee B$ is ranked with B, and
> (c) B is not ranked with A."

His proof here is in two parts. First, he considers the condition where the *prob* of A is not zero, and A is not null. This requires supposing that $AB = F$. From this Jeffrey proceeds to point out that one can show how condition (7.2) (c) above cannot be satisfied if it is assumed that $A \neq 0$, and also that B is ranked with $A \vee B$, thus allowing that conditions (7.2) *(a)* and *(b)* hold.

The second part of his proof supposes that $A = 0$, and that one is to prove that A is null. Proposition B is produced so that $AB = F$, and that the preference-ranking is either (a) or (b) below:

(a)	(b)
A	$B, A \vee B$
$B, A \vee B$	A

With respect to the above, Jeffrey observes that one of the two following choices will always work:

$$B_1 = \bar{A}G \quad \text{or} \quad B_2 = \bar{A}\bar{G}.$$

This is evident since it does not matter whether one selects B_1 or B_2, B will always be incompatible with A, since it was assumed that $AB = F$.[42]

---

[41] ibid., p. 123.
[42] ibid., pp. 124-125.

For at least one choice of B_1 or B_2, the preference-ranking will be either (a) or (b) as shown above. This is illustrated by showing how B_1 and B_2 cannot both be ranked with A. Jeffrey proves this through a series of steps wherein it is noted that given $G = AG \vee \bar{A}G$, and since it is assumed that $prob\ A = 0$, it follows through the expression of the desirability of G that $des\ G = des\ \bar{A}G$, and by parallel argument it is shown that $des\ \bar{G} = des\ \bar{A}\bar{G}$. Applying his analysis of the notions of T and F, he determines the ranking to be as follows:

$$G, \bar{A}G$$
$$T$$
$$\bar{G}, \bar{A}\bar{G}$$

Jeffrey is found to refer to (5-2) as the justification for the above ranking, however it seems that this stems more directly from his definition of T, which occurs in section 5.2, rather than from the Axiom of Desirability.[43]

The above ranking clearly shows that B_1 (or $\bar{A}G$) cannot be ranked with B_2 (or $\bar{A}\bar{G}$). Thus if A is ranked with B_1, then it cannot be ranked with B_2.

By applying the Desirability Axiom one finds that for any choice of B, it is the case that B is ranked with $A \vee B$. What results is that $des\ (A \vee B) = des\ B_1$, which enables one's seeing that B_1 is ranked with $A \vee B$. The same can be shown for choice B_2. Hence it follows that A is null.[44]

The remaining portion of Jeffrey's chapter is devoted to an illustration of a general procedure for comparing the probabilities of A and B, where both are ranked together, as well as a procedure for the measurement of the probabilities of indifferent propositions. In the case of the former, Jeffrey's analysis is straight forward. A "test proposition" C is introduced, so that (7-4): where the existence condition, closure condition and G condition are satisfied, and A and B are ranked together but not with C, so that $AC = BC = F$, and where C is not null, then:

"(a) $prob\ A = prob\ B$ if $A \vee C$ and $B \vee C$ are ranked together;
(b) $prob\ A < prob\ B$ if $A \vee C$ is ranked closer than $B \vee C$ to C;
(c) $prob\ A > prob\ B$ if $B \vee C$ is ranked closer than $A \vee C$ to C."[45]

Jeffrey's proof requires a simple designation for the probabilities of A, B, and C, such that $prob\ A = p$, $prob\ B = q$, and $prob\ C = r$, where $des\ A = des\ B = x$, and $des\ C = y$. There results the ratios of the weights of the two averages of $A \vee$

---
[43] ibid., p. 125.
[44] ibid., p. 125.
[45] ibid., pp. 125-126

C and B v C as: p : r, and q : r. Given that it is not the case that x is equal to y, the determination of which of the values of *des* (A v C) or *des* (B v C) is closer to y is gotten by the ratio of the weights, so that the first value will be either equal to that of the second, as in (a) above, or it will be closer than *des* (B v C) is to y, as in (b) above, or finally it will be further than *des* (B v C) is to y, as in (c) .

Jeffrey's method for measuring the probability of indifferent propositions makes use of definitions (7.3) and (7.4) above. The former allows him to assert that the probability of proposition X is 0 or 1, since either X or $\bar{X}$ is null. However, by assuming that neither X nor $\bar{X}$ is null and that X is ranked with T, Jeffrey proposes to determine the probability of X "to any degree of accuracy."[46]

A few steps in Jeffrey's involved proof can be noted as he proceeds with his analysis.

First, the probability of indifferent proposition X is expressed as $x = prob\ X$. The objective is then to determine to an accuracy of one part in $M = 2^m$ the probability of indifferent proposition X.

Second, the splitting condition is applied to G and $\bar{G}$ m times. This results in having an expression for each n indifferent proposition:

$$I_n = G_n \vee H_n ,$$

where 'H' designates some proposition incompatible with G. Moreover, since the splitting condition is applied here, all disjunctions of G and H are equiprobable, and are ranked together.

In the third step, the proposition:

$$S(n) = I_n, \vee ... \vee I_n$$

is assumed to be indifferent, and as having probability n \1M. In relation to this the definition: $S(0) = F$ is then added. Clearly, it follows that $S(n)$ propositions are formative of an M-step probability scale, from $prob\ S\ (0) =$ to $prob\ S(M) = 1$. It is in contrast to this probability scale that $prob\ X$ is to be measured. This is done by utilizing (7-4), so that $A = X$, and $B = S(n)$, and then deciding which of the following hold: (a) $x = n$\1M, (b) $x = n$\1m, and (c) $x = n$\1M, given $n = 1,...M$-1. Surely, where $n = 0$ one knows that (c) holds, given the characterization of X and $\bar{X}$ as non-null. Also, case (b) holds where $n = M$.

Assuming that case (a) holds for some n, it can be said that

$$x = \frac{n}{M} .$$

---

[46] ibid., pp. 126-127.

On the other hand, if (a) holds for no n, a pair of successive numbers will be generated, e.g., $n, n+1$, so that, case (c) will hold for the first in the pair, and (b) will hold for the second, and *prob X* will fall somewhere within *prob S(n)* to *ProbS(n+1)*, inclusively.

In the final step of his proof, Jeffrey considers fulfilling conditions of applicability for (7-4), *whenever X is indifferent, X and $\bar{X}$ are both not null*, and given that $n = 1,...,M\text{-}1$. The above presuppositions allow that $\bar{X}$ be indifferent as well. In essence, Jeffrey's proof follows along the lines of showing how no particular pair of the kind: GX, $G\bar{X}$, $\bar{G}X$, and $\bar{G}\bar{X}$ has a disjunction which is null, and which also does not violate some foregoing condition for X and G.[47]

Jeffrey summarizes the results of his seventh chapter by observing that where the closure, G, and splitting conditions are satisfied, then: (a) the preference-ranking *uniquely* determines the ratio of probabilities of any two propositions which are ranked together, and consequently, (b) prefence-ranking also *uniquely* determines the ratio of probabilities of all indifferent propositions.

The subsequent chapter Jeffrey devotes to proving a uniqueness theorem for his theory, making extensive use of the summary he presents above. Detailed elaboration of this aspect of his work hardly seems necessary, since what he does is simply to show by 7.1 how the totality of all preference-preserving probability and desirability transformations are those discussed in Chapter 6. Similarly, Bolker's axioms, as these pertain to the *existence problem*, are only briefly discussed by Jeffrey in Chapter 9., and are taken as lying far beyond the scope of the relatively simple mathematics which has underlined his theory thus far.

Jeffrey's work in *The Logic of Decision* is held by him to be extended by his more recent analysis in the essay "Preference Among Preferences." The difference between both works is that "preference" is treated by him as a two-term relation in the book, whereas in the article "preference" is seen as a proposition *about* a proposition, which necessitates the introduction of a new connective. What he hopes to bring off in this later work is a formalization of the complex modality which involves successive orders of preference.

Whether Jeffrey can defend his claim that there is a continuity between his theory of preference and his analysis of second-order preferences remains to be discussed. What needs to be assessed first is Jeffrey's central position in "Preferences Among Preferences."

---

[47] ibid., pp.127-129.

Section 2. *Jeffrey's Formal Analysis of Orders of Preference*

Jeffrey's work in the essay "Preferences Among Preferences" differs from those efforts of the past which sought to give logics for purely the evaluative or cognitive aspects of first-order preference.[48] For he sees order among preferences, which is open to formal characterization by way of strict entailment. In essence, he is providing a logical foundation for a number of "real distinctions," drawn through *abstraction*, from the *significance* of certain preference expressions in ordinary discourse.[49] It is important to note that Jeffrey is not offering a theory of preference, in the same sense that he is endeavoring to account for the preconditions which must give rise to the situations involving preferences. Rather, he considers preferences as given *in discourse*, and then proceeds to formalize their implicational bond.

Jeffrey states that the successive orders of preference are a result of the unique fact that humans are self-conscious about their preferences, and are usually — though not always — able to understand their full ramifications. Thus, one requires a means by which to express the strongly implicatory character between preferences. Jeffrey claims that preference should be expressed as the *whole* complex expression: "X pref Y," meaning that X is preferred to Y. The preference expression here is not to be taken as two different things to be connected by the symbol for preference. Rather, the expression should be taken as *the entailment* within a connection between two options, X and Y, where the former is preferred to the latter.[50]

His example of a higher-order preference is as follows:

1. ((-S pref S) pref (S pref -S)) pref (S pref -S). 1. expresses the relation that if some individual prefers not smoking, i.e., -S, to smoking, S, then he could prefer to smoke, provided that whenever he prefers to abstain, he could also prefer to smoke. More precisely, 1. is saying that if one prefers not smoking, -S, to smoking, S, then he could not possibly be preferring not smoking, (-S pref S), unless at the same time he *could also* prefer to smoke, (S pref -S). For in preferring to abstain over smoking, ((-S pref S) pref (Spref -S)), he could also (at the same time) prefer to smoke, (S pref -S).[51]

---

[48] Rescher, Nicholas, *The Logic of Decision and Action*, University of Pittsburgh Press, 1967, pp. 39-40, also Richard M. Martin's *Intension and Decision*, (New Jersey: Englewood Cliffs 1963), Chapter 2, on Preference," and *Belief, Existence and Meaning*, New York University Press, 1969, pp. 258-259.

[49] Jeffrey, Richard C., "Preferences Among Preferences," p. 380.

[50] ibid., p. 378, and p. 383.

[51] ibid., p. 378.

The initial move of Jeffrey's approach is to say that in order for an expression to count as an expression of preference, there must be two *relata* which flank the preference primitive, and that these *relata* are in themselves sentential expressions of propositions concerning the opposing courses of action which the agent *believes* he can take. In the ideal expression of preference, there are always two possible options from which the agent can choose, granted that there is the absence of compulsion either from habit or external pressure. Thus, the symbol for preference, 'pref', comes to convey the idea of connecting the enactments of a sentence expressing a course of action with one's nonenactment of a sentence expressing a contrary course of action. Moreover, the expression of preference also conveys the idea that the course of action alluded to at the left of the preference connective is preferred to that on the right, so to speak, so that the latter course of action can *never possibly* be preferred over the former state.[52]

It is important to emphasize that the two options which are equally available to the agent are not changed by saying that one of them is preferred over the other. In Jeffrey's sense, to prefer an option does not mean that therefore the other option, to the right of the connective, is in principle no longer an option that the agent can consider as a course of action he could take. Rather, Jeffrey is saying that where one of the options is preferred, its contrary could not meaningfully "be preferred," though the latter is still an option. Surely, these options must retain their inherent viability if preference is going to convey its characteristic feature of uncoerced choice. Consequently, Jeffrey's sensitivity to the "modal involvement" of successive orders of preferences is keenly manifested.

Perhaps one of the leading points to question at this juncture is the way in which Jeffrey speaks of the "enactment of a proposition," where he talks of how one acts upon a preference. What does it mean to say that one *enacts* a proposition? How does Jeffrey conceive of such enactment?

Granted that this is a fair assessment of Jeffrey's conception of the proposition, a number of difficulties are seen to arise. First, it is crucial to recall that for Jeffrey it is the whole preferential expression, with its constitutive *relata*. which somehow reflects the entailment of the preferential state of affairs. However, the issue of what exactly is being reflected by such an expression is quite formidable. For Jeffrey presents his conception of the preference-relation as where the individual prefers the option to the left of the preferring relation.

---

[52] ibid., pp. 382-383.

Hence, it makes sense to say that the agent prefers X, where X is some possible and actualizable state of affairs *of* the world. However, how can it be said as well that the same preference expression *also reflects* the option which he does *not* prefer? Though in Jeffrean terms the latter option is a possible course of action the agent could choose in itself, this is a different matter from the broader issue of what the *entire* preference expression can be said to genuinely refer to. For the option which in not preferred is, in the context of some particular preference expression, never a state of affairs which will be brought about. Yet, here one may ask how the entire preference expression can be *both* expressive of a possible state of affairs (the preferred or the state of affairs to be enacted), as well as the condition which will not be brought about by the subject (i.e., the state which is not preferred)? In Jeffrey's analysis the preference expression reflects or signifies both a possible and actualizable situation, and on the other hand, a possible though nonactualizable state of affairs. Though both options are understood to be possible, the connective 'pref' serves to render the non-preferred option unactualizable within the context of preferring. In brief, within the context of a preference expression, the second option is possible though *also* unactualizable. One wonders here whether Jeffrey's ontology with respect to the reference of his conception of preference expression is any clearer than the confused ontology he charges Ramsey with, where the latter is seen to treat propositions and gambles on an equal footing.

A number of additional problems arise from what Jeffrey has presented thus far. First is the issue of whether what is not preferred need necessarily be always the contrary of what is preferred. The point being that perhaps the preference connective Jeffrey defines need not express exclusivity in order for it to be expressive of preference in "ordinary talk," which is Jeffrey's objective in presenting his inquiry into preferences. Furthermore, there is one further problem with Jeffrey's analysis, namely that where one introduces an agent's acting to bring about state of affairs, say x, then probability enters into the discussion, with respect to one's success in achieving one's goal. Thus, Jeffrey's allusion to possible states is found to involve the role of probabilities when the full sense of his preference expression is unfolded. Consequently, his expressions of preference do not deal *solely* with the ontology of the possible, but they also intimately involve the conceptually different modality of the probable.

Difficulties emerge as well where one reflects upon the manner by which Jeffrey speaks about the "enactment" of a preference. In the course of his presentation, Jeffrey interchanges the word "proposition" for the word "sentence," so that to enact a proposition is also taken to mean that one makes some corresponding sentence true.[53] Again, there is little effort by Jeffrey to draw any of the fine distinctions between proposition and sentence. Thus, one is not sure as to how he should interpret the *relata* which flank the preference-relation. Jeffrey does not want to go in the direction of saying that the *relata* express sentences about existent entities in the physical world. For then they would involve the actual and not necessarily the enactable. On the otherhand, he could not interpret propositions as the *meanings* of the *relata*. For meanings as cognitive entities resulting from reflections within consciousness, are not clearly seen to be states of affairs which can be purposefully enacted. At least it seems wanting to say that one enacts meanings, assuming in some way that meanings are simply there, and that someone acts upon them. Moreover, to suggest that Jeffrey intended that propositions were to be construed as meanings clashes with his manner of interchanging in the body of his text propositions for sentences. The interchange would be impossible to support, if one where to take his reference to meanings in the sense of "belonging to" sentences. On such a view a sentence would become more of a purely syntactical entity, rather than a semantical one — as Jeffrey must have intended by his many references to the enactment of sentences.

Perhaps, as Jeffrey casually suggests, the *relata* could be interpreted as imperatives.[54] In this way, at least, one could make sense of the notion that in acting upon a preference one must in some sense "enact" one of the *relata* which flank the relational term. However, if these *relata* are to be construed as imperatives, then who is giving them to whom? Since preference is usually considered as free personal choice, the imperatives which would then flank the preference-relation must also be said to be self-reflexive, in some sense. This is to say that the person doing the preferring must be commanding himself. Hence, on this interpretation one is faced with the odd situation where one is both the commander of and the respondent to the same command. Moreover, interpreting the *relata* as imperatives detracts from Jeffrey's recognition of the requirement that uncoerced choice is an essential aspect of the preference expression. Surely, imperatives suggest a compelling and necessary mode of response, rather than an invitation to free choice. Finally, the formidable question arises as to the way one is to explain the *relatum* which expresses that which is not preferred. In other words, there seems to be a problem as to how, or

---

[53] ibid., pp. 382-383.
[54] ibid., p. 388.

under what conditions, one would give a command for that which would not be enacted (i.e., the nonpreferred).

Thus far it can be said that Jeffrey's enterprise has been arrested somewhat by the issues raised concerning the extension of preference expression conceived in his manner. For there seems to be a conflict between the claim that preference must reflect two mutually exclusive though possible states of affairs, and the claim that contextually (i.e., intentionally) that which is preferred is something more than *just* a possible option expressed by an uttered sentence. Curiously, a tension emerges between the extension or reference of Jeffrey's view of preference, and the intention of this concept, seen as the sum total of its possible connotations.

Allied with the questions regarding the ontological ramifications of Jeffrey's notion of preference is the issue of how one is to justify the allusion to Lewis' idea of strict implication within the context at hand. Surely, understanding that preference expressions reflect possible states of affairs is fundamental to any clarification of the propriety of referring to a modal connective like strict implication in a semantics of preference. Historically, Lewis introduced this notion so as to restrict the formal sense of the "if..., then ...," connective, wherefore it could be said that it is impossible, and not just false, that p implies q, where p is true and q is false.[55] Consequently, strict implication expresses an "absolute" necessity, which emerges from the very meaning of the terms it connects. Jeffrey refers to it analogously, stating that 'pref' performs a similar function in expressing the strong implicative which he sees operating within preference. The latter nontruth-function he sees as also animated by the full spectrum of human motivation: desire, want, wishing, etc.[56] However, it is unclear whether Jeffrey is justified in alluding to a connective expressing logical necessity with a modal context which was seen to deal with the physically possible.

The fundamental question here is whether the exclusivity (so-called) of preferring arises from factors *inherent* in the activity of preferring itself, or whether it is really the result of investigations into the contextual periphery of preference. Jeffrey operates on the supposition that one sees, in an apparently introspective manner, that inherently within the act of preference there is operating the exclusiveness of what is preferred from its contrary. However, he

---

[55] Edwards, Paul, *The Encyclopedia of Philosophy*, Vol. 5, (NewYork: Macmillan Company and the Free press, 1967), pp. 5-6.
[56] Jeffrey, Richard C. "Preferences Among Preferences," pp. 386-387.

may be trying to bring under logical rigor an implication which is not inherent in the preference itself as expressed in the direct voice. This is to say that the individual who is preferring does not (off-hand) seem to be consciously preferring not to prefer the contrary of that which he prefers. More simply, it cannot be said that from an ordinary language point of view, and Jeffrey gears his investigation to be sensitive to this viewpoint, that in preferring one is also aware of two mutually exclusive states of affairs, one of which he prefers. Rather, preference on the face of it appears to be like a dispositional attitude, much in the same way that assertion is a dispositional attitude. Similarly to assertion, preference does not of itself, as expressed in the direct voice, convey any information about a contrary state of affairs.[57] One becomes aware of the options which a preference creates only *after* testing comes into play, and the preferring individual reports through his own utterances what he does or does not favor, relative to the background of his knowledge of his preference. In essence, preference alone is merely the expression of what one would like seeing or happening. It requires a *further step* of investigation to justify saying that one would oppose such and such a state of affairs because he prefers some other state of affairs. One cannot proceed *a priori*, as Jeffrey seems to be doing, to state uncompromisingly that the agent does not prefer the contrary of what he prefers without introducing qualifying conditions. Often the implications of a preference are a complex state of affairs which cannot be expressed simply as the complement of that which is preferred.

It is questionable then whether Jeffrey is justified in presenting the preference expression in terms of two *logically* exclusive conditions. His claim that he sees the full spectrum of human motivation involved in such expression suggests that he is viewing preferences subjectively, as dispositional attitudes. Yet, by introducing two exclusive aspects in preference, he is also suggesting a deeper level of analysis, one which presupposes objective testing and inductive generalizations. The point here is simply that Jeffrey is allowing too great a degree of complexity for a dispositional attitude. Interestingly, one may note how Quine aptly recognized that dispositional propositions are referentially opaque because they prohibit, through their failure of extensionality, codesignative terms from operating.[58]

In view of the inductive foundation which must be realized in the analysis of preference, the question emerges of whether or not one can allude to

---

[57] Geach, Peter, "Assertion," p. 254.
[58] Quine, Willard van O.,, *Word and Object*, (Massachusetts: MIT Press, 1964), pp. 150-151.

strict implication to explain exclusivity between options of preference. More succinctly, can one consistently speak of the logically impossible in reference to the expression of preferences? This question is predicated on the observation that in order to speak of any sort of implication among preferences, one must consider the action(s) which the preferred object comes to require, and the effect of that action in the agent's physical environment. Though the totality of these effects are beyond complete consideration, still a minimal amount of attention should be given to the role they must play in the ordering and the interassociation of preferences. Jeffrey says very little about the contingencies which surrounds the effects of action when he considers the nature of the implication between the *relata* of a preference expression. As a result, his presentation tends to suggest an absolute exclusivity between the *relata*, which is insensitive to the reality of the situation at hand. Surely, options which were once thought to be impossible (physically) to co-exist are no longer thought to be so related because of innovative technical advances. One can easily point to options which were once thought to be contrary, but which have since become compatible. For example, preferring to live in an atmospherically dry region is no longer the contrary of living in a fruit-yielding area. One sees here how incongruous it is to attempt to impose logical necessity upon a relation of options which come to constitute the expression of preference. Such an effort misses the contingency which surrounds the conditions of preference.[59]

The latter point brings into view an additional facet of the concept of preference. This is to say that there is an unavoidable element of indeterminacy involved with whether what is preferred *can* occur. In the fuller sense of preference which is being considered here, one must take into account the inductive inference of whether or not some state of affairs can occur in the physical realm, if certain action is taken. Preference expressions seem to be posited on the presupposition that "all things being equal," such and such is to occur. It would make little sense to utter a preference without the concomitant understanding that it may or may not come about. Jeffrey seems to be recognizing the same point where he introduces the notion of enactment to explain how one acts upon a preference. Though Jeffrey's insight was seen to lack clarity upon examination, it still reflects the need to introduce some aspect of judgment about action when dealing with preferences. Indeed, one has here a basic

---

[59] Jeffrey's notion of preference presumes the involvement of wants, desires, and wishes. This, in turn, makes an objective analysis of preferences difficult to secure. It is the author's opinion that preference should have a logic distinctly its own, separate from considerations of wishing and desiring.

means of distinguishing between wishing and preferring. In the former case, the object of one's wish coming true or not coming true is not an integrated part of the expression of the wish. One may wish for something while knowing very well that no matter what happens there is not the ghost of a chance that his wish will come true. Preference, on the other hand, is predicated on the probability that there is free choice amongst alternatives which some agent can take. To introduce an expression of preferring in a situation where there is no possibility for that which is preferred to come about is somehow to have an *idle* preference, or a situation where there are not free options and hence, no true preference.

Thus, the inductive element appears to be an essential aspect of a preference expression, where the latter is considered in the fuller sense of its empirical ramifications. For this reason it becomes difficult to sustain the claim that the intuitively understood ramifications of preferences in spoken discourse can be connected by means of a notion allied to strict implication, which reflects logical impossibility.

Already within the considerations dealing with the applicability of the notion of strict implication the issue of what one is said to know when he expresses a preference arises. For it was seen that the expanded sense of preference required considering empirical states of affairs, and actions which operate in bringing about these states of affairs. Thus, the cognitive element invariably comes into view when considering the full ramifications of preference. Jeffrey himself makes a strong effort to accommodate this aspect of the investigation where he notes the important role of "recognition" in determining whether the option which helps to compose an expression of preference is or is not free.[60] He encounters difficulties, however, in explaining just how "recognition" is to be made explicit. One approach he proposes is to say that recognition involves a "full belief in a truth" as expressed by a sentence concerning preference. This state of belief is to be conjoined in some manner with what he terms the agent's "judgmental probability function." The latter appears to be the agent's assessment of whether his preference will come about as he expects. He suggests that this is measurable since he assigns a special term 'P' to range over the value of this function. He seeks to secure further the tenability of his logicized semantics by adding terms for the supporting factors of "expectancy" E, and for "utility" u. The two latter are deemed necessary in presenting a complete description of higher-orders of preferences.[61]

---

[60] Jeffrey, Richard C., "Preferences Among Preferences," p. 386.
[61] ibid., pp. 388-389.

These factors which he introduces: "full belief," "judgmental probability function," "expectancy," and "utility," all presuppose some degree of knowledge concerning the object and the circumstance of the preference. However, to merely state the need for introducing these factors without discussing the epistemic conditions they involve does not lead toward a useful description of the notion of preference. The point is not that Jeffrey is wrong to note the importance of these factors in explaining what preferences involve. Rather, his error is in neglecting to explain the nature of the knowledge claims which come to constitute the values for the functions of "full" belief, probability, utility, and expectancy. If one were to press Jeffrey on what is meant by "full" belief in a truth, he would find a very shallow explanation of what is involved. What is a "full" belief? How is one to draw the line between a full belief and a partial one? Jeffrey does not go into what is involved here, just as he does not explain what he means by "judgmental" probability function. The latter appears to be not merely an objective factor which can be statistically determined, but it is somehow linked to a subjective factor of belief which seems to depend upon an introspective awareness which does not have a public access. This dearth of exegesis only underscores the shortcomings of the intuitionistic mode of analysis which Jeffrey chooses to adopt. Namely, it commits him to entities of an obscure nature, which are of dubious interest to investigators in the social sciences, e.g., economics, sociology, etc.

Jeffrey's neglect of the epistemological aspect of preference makes his analysis unclear. This is to say that one is unsure of the standpoint from which Jeffrey is performing his investigation. If he is approaching the description of "preferences among preferences" from the standpoint of one's conscious knowing of the ramifications of his preferring, then his analysis is tantamount to a phenomenological observing of different levels of mental reflection. This makes Jeffrey omniscient in that it would be supposing that he knows other minds. On the other hand, if he is talking about the "preferences among preferences" from the viewpoint of an external observer, who is considering the external reports of what is being preferred, then he should be allowing for more of the contingencies which operate in the determination of states of affairs in the physical realm. Yet, his presentation of the constituents of preference in terms of logical exclusivity has shown that he is unwilling to consider the investigation of preferences on any other than an intuitionistic level. Thus, he is found to be snared between an intuitionist method of analysis, and the temporal requirements which speaking clearly about the nature of preference necessitates.

The ambiguity surrounding the direction of Jeffrey's analysis leads toward a serious challenge of the meaningfulness of speaking of "preferences among preferences." It was found that he sees the virtue of introducing the "pref" connective in its being able to express the entailment of self-reflexive exclusivity which a single expression of preference involves at deeper levels. Thus, the orders of preference which he sees are indicative of the telescoped meaning of preference expressions. Yet because of the ambiguity surrounding the nature of the type of analysis Jeffrey engages in, i.e., whether phenomenological or empirical, it is very much open to question whether anything significant is being talked of when he concentrates upon the supposed self-reflexive depth of expressions of preference.

To illustrate the above point one can refer again to Jeffrey's analysis of one's preferring to abstain from smoking. In such a situation one has two options, to smoke and not to smoke, and the individual prefers the latter to the former. He goes on further to state that to prefer to abstain can be analyzed further as meaning that one prefers preferring not to smoke to preferring to smoke.[62] Presumably, one can go on to talk of the preferring of one's preferring, etc., etc. The question thus becomes: is anything being meaningful added to the first-order expression of preference by introducing the regressive "preferring" locution? Quite simply, it seems that where one is said to prefer preferring to abstain, he is essentially saying that he prefers to abstain. It seems that Jeffrey is unnecessarily complicating the sense of the first-order expression. Though "preferring" can be taken as an object of preference where one prefers to prefer (as an option) rather than not to do any preferring at all, this matter of "preferring to prefer," of which Jeffrey endeavors to articulate a logic, is different from that of the case where preference is considered as an option. The former situation, where to prefer is taken as an option, has an identifiable context, where determinations can be made concerning whether or not it can indeed be said that one has the option of preferring. In the latter case, however, Jeffrey comes to recognize that he is dealing with attitudinal states, which are in some sense packed into first-order preference expressions.[63]

Insofar as Jeffrey is concerned to present a logicized semantics of the attitudinal aspects of preference, he is directly concerned with that facet of preference dealing with the private state of awareness, which is distinctive

---

[62] ibid., p. 390.
[63] ibid., p. 390.

with the case of preference. However, as Wittgenstein points out, private experiences cannot be objectified and categorized because their exhibition is not open to public inspection. Though one often speaks as if he is communicating to others facts *about* his mental processes, upon closer analysis one finds that talk about mental states is a fiction, since nothing can be meaningfully communicated about that which has no public reference. Returning to Jeffrey then, one finds the hierarchies of preference he sees in this attitudinal states do not appear to reflect any substantive demarcations, contrary to what he attempts to suggest. Consequently, the whole enterprise of presenting a *formal* analysis of these attitudinal states is open to question, and the claim that lower-orders of preferences can become arguments for expressions of higher-orders of preferences seems equally dubious. Basically, the referential opacity of attitudinal dispositions like preference spoken of above prohibits the determination of the supposed depth which Jeffrey believes he can intuit in such expressions. It is for this reason that one can say that the "preferring" locution does not seem to add anything meaningful to first-order expressions of preference.

Moreover, the vertical ordering of the hierarchies of preferences by Jeffrey also takes on a highly subjective character, and thus it cannot be given the public justification it requires. For example, toward the end of the paper Jeffrey presents the following as a consistent ordering of preferences:[64]

$$S$$
$$-(S\ pref\ -S)$$
$$S\ pref\ -S$$
$$-S$$

Jeffrey observes that the top most preference S (smoking) is ranked above the second -(S pref -S), i.e., (not preferring to smoke) because it manifests "...the best of the (recognized) options." This act of recognition is based fundamentally upon the agent's "...intensity of ...preference." Evidently, each of the expressions within the above ranking constitute options which are open to the agent, but with the added factor that they each differ from each other relative to the strength of the emotive intensity with which they are preferred.[65] Hence, moving downward from the top, the intensity of preferring is lessened relative to a particular agent at some particular time. Apparently, the above ordering is to be explained where the case of *actually* smoking has a greater preferential enticement than that of preferring to abstain from smoking. The latter is in turn

---

[64] ibid., pp. 390-391.
[65] ibid., p. 390.

more intense as a preference than the preference for smoking (S pref -S), since the former should be pursued as the "best of the (recognized) options." The last option, not smoking (-S), has zero intensity, apparently because it is a state where no attitudinal consciousness is operating, e.g., being sound asleep.

The so-called ordering of different expressions of preference again reveals the obscure basis upon which Jeffrey chooses to make presumed logical distinctions. The levels of priorities which he sees in preferring are not clearly independent from the empirical factors, which above were seen to be unavoidable in considering a full characterization of preference. For example, the top most preference, smoking, seems to entail with it the habit of smoking, which in turn puts it on a higher level than abstaining because of the former's greater intensity. Thus, the difference between the preferences actually turns out to be based upon a physical habituation, rather than upon any logical necessity. Interestingly, the difference between the second preference, abstaining, and the third, smoking, is based upon a "should" which supposedly reflects the "best of the (recognized) options." Here again it can be said that the distinction between the two preferences (the second and the third) is based upon the empirical requirement of what one should do for presumably the protection of his health. For it appears that one interprets Jeffrey as saying that he should not smoke, or that he should abstain from smoking, because from the point of view of one's well-being, it is not healthy to do so. The distinction between the two preferences is once more founded not upon a logical necessity, but upon an observation of physical phenomena. Finally, the third preference is distinguished from the fourth, nonsmoking, again upon a degree of intensity which arises from empirical factors. For it appears that the state of a conscious choice between two alternatives, to smoke or not to smoke, is in some sense a more vivid (i.e., intense) state than the very open and ambiguous one of nonsmoking. Thus, the distinction is once more being made on the basis of an observed state of affairs, which is an experiential generalization.

In view of the above, one may well question the meaning of the ordering Jeffrey endeavors to logicize as not based upon any consistent reasoning. Again, one may note how his analysis becomes burdened by the empirical requirements of that which is being analyzed, and how the limited dimensions of an intuitionistic approach seems unable to cope with the full depth of the notion of preference.

The intuitionism implicit in the Jeffrean approach makes problematic the ontology of preference expressions, and his manner of excluding the praxiological facets of preference leads to serious doubts about the utility of the notion of strict implication within a context of preference. Furthermore, it was seen that basic epistemic requirements for preference were treated incompletely, producing less than a clear resolution as to the perspective from which he is performing his analysis. Finally, the value of having a logic of preferences *among* preferences is debatable given the difficulty of showing whether anything is gained by introducing a manner of speaking which supposes various levels within the activity of preferring (e.g., preferring to prefer).

The difficulties which are seen in Jeffrey's presentation are of value only insofar as they can instruct one as to the problems one encounters in analyzing the subject at hand. Thus, it appears to pervade Jeffrey's pioneering effort that one cannot do justice to the complexity of the notion of preference by pursuing it purely from an intentionalist viewpoint. For preference seems to warrant another approach, rather than the intuitive approach suggested by Jeffrey. In effect, it was seen that he had to resort to the extensional aspects of preference, where he spoke of "enactment of a proposition," "judgmental probability," "expectancy," etc. The latter all come to involve the extrinsic aspects of preference, and are best explained denotatively. His resorting to the extension of preference appears to conflict with his claimed intuitionistic analysis.

PART TWO

The Subjectivist Approach Toward the Formalization of Preferences

Chapter 5.

Soren Hallden's "Puristic" Logic of the Better and Same

In considering Hallden's work, it is vitally important to keep clear that he is not presenting a "logic of preference" in *On the Logic of 'Better'*[1]. This is an especially difficult point to bring across since not only do most researchers in the field of preferential logic cite his contribution in one connection or another, but Hallden himself is found to allude to the betterness of p to q as meaning in a normative sense: "p is to be preferred to q."[2] However, in his most recent book, *The Foundation of Decision Logic*, Hallden is careful to distinguish between preference *per se*, as involving "empirical" relations, and betterness as dealing with evaluative recommendations.[3] It is the latter which are submitted to formal analysis in the first work mentioned above. In retrospect, the evidence suggests that Hallden's realization of the necessity to distinguish between discourse about preferences from discourse dealing with the better arose subsequent to Henrik von Wright's seminal work, *The Logic of Preference*.[4] In 1963 von Wright characterized the difference between *deontological* concepts involving norms, e.g., duty, command, permission, etc., *axiological* concepts involving value, e.g., good, bad, and better, and *anthropological* concepts involving notions relating to need and want, decision and choice, etc. However, von Wright was also quick to point out that these groupings are not to be studied in total isolation of each other, since invariably one finds an overlapping between evaluative, normative and pragmatic concepts.[5]

Thus, to the question of why Hallden's work deserves consideration here, a number of replies are warranted. Essentially, one could argue that in view of von Wright's important insights, it is very much an open question whether one can have a logic of the better and the same which is "puristic" or somehow neutral in the sense that it is totally divorced from any possible con-

---

[1] Hallden, Soren, *On the Logic of 'Better'*, Library of Theoria, Upsala, 1957.
[2] ibid., p. 12.
[3] ibid., *The Foundations of Decision Logic*, Library of Theoria, No. 14, Lund, 1980, p.11.
[4] von Wright, Georg Henrik, *The Logic of Preference*, Edinburgh University Press, 1963.
[5] ibid., p. 7.

sideration regarding practical action and therefore of preference; though such a puristic logic is taken as the major objective in *On the Logic of 'Better'*.[6] As will be seen, whenever there is a question as to what Hallden means by his purely intuitive rendition of the relations of betterness and sameness, the issue turns into one involving the course of action one would take in preferring one state of affairs as opposed to some other. At various points the opacity of Hallden's logic is found to be due to a dearth of reference to praxiological factors. This, however, underscores the difficulty of keeping separate the study of the formal properties of evaluative notions such as the better from an inquiry into the formal properties of the notion of preference, for example. Nonetheless, understanding Hallden's logic requires that one allude to some notion of preference as a means of clarifying what it means to say that one state of affairs is better than another.

Also in the context of the evolution of the literature in this area, Hallden's work is seen by von Wright as related to his own logic of preference. Von Wright claims that Hallden's attempt, in its concentration upon relations between states, resembles his own work on preferential "states." Moreover, some of Hallden's results are held to be "valid" within von Wright's own formal theory.[7] Thus, Hallden's work in 1957 becomes an important key in understanding the direction of von Wright's effort which attempted to face the issue of the nature of the logic of preference more directly than Hallden.

Soren Hallden prefaces the introduction to his logic by examining how evaluative terms like "the good," "better," and "same," operate in ordinary discourse. While exhibiting a sensitivity for the way such expressions occur in everyday parlance, his own views on the better come to reflect what he interprets as being a "purer" conception of logic of such discourse. In pursuit of "clarity" he spurns any reliance upon the meandering nuances of ordinary discourse. Rather, Hallden introduces his remarks by observing that spoken discourse exemplifies a deficiency which his logico-analytical approach will attempt to rectify.[8] His recognition of how ambiguity permeates natural language reflects as well a common skeptical attitude toward language becoming a reliable source of philosophical information, shared by Davidson and others in earlier attempts. However, the important difference is that Hallden, unlike previous investigators, does not reject as virtually useless any study of ordinary discourse for the purpose of gaining philosophical insight. On the

---

[6] Hallden, Soren, *On the Logic of 'Better'*, p. 12.
[7] von Wright, Georg Henrik, *The Logic of Preference*, p. 18.
[8] Hallden, Sorren, *On the Logic of 'Better'*, pp. 12-15.

contrary, he employs the examples from natural language as a foil by which to vindicate his own perception of what a correct logic of the better should be. In this way he believes his work has the advantage of providing a way of illuminating the subtle relations holding between value concepts, as *ordinarily* understood.

Hallden presents two theories of the "better," both of which he admits to being deficient in various ways. Yet apart from their limitations, Hallden's work signals a serious effort toward achievement a formalization of "better" within a fully articulated logico-philosophical context. In this, his effort constitutes a further point of departure from that of Davidson and his associates. It will be recalled that the latter concentrated on the capacity of probability theory, employed in the study of economics, to elucidate the logic of preference within a philosophical domain. Though Hallden's effort is short on discussing this aspect of the matter, his is still an attempt which focuses mainly on the power of formal logic to clarify an evaluative concept. As such, Hallden's work constitutes the first attempt to evolve a completely formal theory of the better and same within a homogeneous realm, i.e., where the tools of investigation, as well as the object investigated, are within the same area of inquiry.

However, apart from his innovation, it is significant to deliberate upon Hallden's basic claim that he is solely interested in the formal "order of value." The difficulty which his enterprise elicits results from the juxtaposition of these two very terms, namely, "order" and "value". Whereas the former clearly provides for the possibility of considering the "formal" aspects of relations, the latter, i.e., "value", ordinarily appears to be less formal, since it is a context-dependent notion. Hallden contends, however, that value can be treated in a totally context-free mode, whereby one can study the formal structure of value relations. Yet, what sense does it make to say that one can consider "value" from a purely formal perspective? In a very broad sense one could talk of value as the premium one places on possessing something, or perhaps as the specific degree of approbation or lack of same given to a particular mode of action or state of affairs. However generally one may conceive of value, there is always a context of human concern which seems to lend it intelligibility. In this, one is reminded of von Wright's incisive conclusion that one cannot in practice separate the study of deontology from that of axiology and anthropology. The study of the "order of value" is surely well within the area of axiology. However, without the added considerations dealing with what is the case in a context of human interaction, the study of the "order of value" is vacuous. The philosophically naive position to which Hallden is willing to subscribe, renders the object of his investigation difficult to explicate. Somehow the study of the formal syntax of the better and same is found to be useless in penetrating and explicating the semantic and pragmatic relations, which constitute the essence of what it means to say that x is better than y, or that therefore x is preferred to y.

I. *Conceptual Preliminaries*

The foundation of both theories of the better Hallden presents involve the two relational connectives, 'B' and 'S', expressing the notions of "better than" and "same as" respectively. Hence, 'pBq' expresses the sentence that p is preferred to q, where 'p' is to be read as any variable which expresses the possibility of p, and 'q' expresses the possibility of q, when correctly interpreted. Thus, p and q are to be taken as propositions expressing different possibilities, and consequently, 'p' and 'q' are propositional variables. For Hallden, a proposition generally is any expression which can be said to be true or false. The expression 'pSq' means that possibility p is the same (or equal to) the truth value of q. Hallden notes that 'pBq' and 'pSq' are propositional expressions of comparison between any two possible world states, p and q, to which a value of truth or falsity can be attached.[9]

Though his use of propositional variables provides little out of the ordinary, the meaning he gives to both 'B' and 'S' requires special attention because of its innovative quality. For neither of these relational terms is meant to be taken as conveying any of the wide variety of meaning which "betterness" and "sameness" usually convey in ordinary discourse. To secure a "stricter and purer" sense of these terms, Hallden stipulates first that they will not be used to suggest temporal specificity, e.g., past, present or future. Second, neither 'p' nor 'q', nor any proposition derived from them, will be taken as *commonly asserted* or *implied* by sentences such as 'pBq' and 'pSq'. This specific limitation is designed to prevent the possibility of expressions within the theory from implying propositions in a nonformal manner, as for example, where one expression is said to imply another in everyday discourse. Again, one sees here how Hallden requires that only *logical* connections between propositions be allowed. Third, 'pBq' and 'pSq' will be used in such a way that no modal relations are implied between p and q. This by-passes any possibility for an open-ended interpretation, which usually results when working with locutions involving the subjunctive mood. Hallden carefully avoids expressions prefaced by "... it would have been. ...", which defy interpretation in terms of the relation of implication found in formal logic. Fourth, neither 'pBq' nor 'pSq' will be taken in such a way as to suggest that something is being asserted about

---
[9] ibid., p. 12.

the individual goodness or badness of p or q. This stipulation is used by Hallden to preserve the purely formal ordering which these sentences are meant to express, and thus avoids any reference to value judgments of a nonlogical character. Finally, no evaluation is presumed as a *prerequisite* for expressions such as 'pBq' and 'pSq'. With this final point, Hallden seeks to avoid the confusion of trying to fathom an infinite regress of possible justification for these two fundamental expressions. Having illustrated the many ways in which 'B' and 'S' will depart from the ordinary sense of "betterness" and "sameness," Hallden closes his attack upon the reliability of ordinary language as a source of philosophical enlightenment.[10]

His logic of the better has importance for him only insofar as it can help clarify ideas in investigations involving theories of value. With respect to the latter, he goes on to observe that the "formal" relations which his theory deals with, i.e., what 'B' and 'S' express as relational components, are neither "intrinsic" nor necessarily "all-pervasive," but neutral. This means that the sentential expressions pBq and pSq do not reflect judgments dealing with the value of things *in themselves*, (intrinsic judgments), nor do they necessarily reflect *total* calculations, which take into account all the possible long range effects of actions. Rather, the relations which will be formalized are of a "neutral" kind, where there is no statement concerning the supporting grounds for the comparisons themselves, only that of their logical consistency. The example Hallden gives is where one would say: "It is better to have a Swedish army than not," and then does not give any supporting reasons for the evaluative judgment he is making.[11] This, for Hallden, is the sense of having a "neutral" comparison. Hence 'B' and 'S' in sentences pBq and pSq will be taken as expressing neutral comparisons of this kind.

In addition to his insistence on a contextual neutrality for better and same expressions, Hallden is also careful to dissociate valid "ethical" inference within deontic logic, from logical validity *per se*. He insists upon a difference between what he terms the "ethical validity" common to expressions dealing with the better and logical validity. For example, ordinarily a logician would not be interested in investigating the validity of the expression: 'pBq ≡ (pBq)B-(pBq)'. The latter is not exclusively an assertion of formal logic, where the analysis of the formal properties of expressions is the main objective. Rather, this expression is one that is accepted as universally valid from an *ethical* point of

---

[10] ibid., pp. 13-14.
[11] ibid., pp. 15-16.

view, apart from its logical truth.[12] This distinction between logical validity and ethical validity is apparently of great importance for Hallden, though as is the case so often in his writings, there is precious little by way of explanation on just how the "ethically convincing" is to be construed.

Keeping in view the perspective of deontic logic from which he is operating, the expression: 'pBq ≡ (pBq)B-(pBq)' is comparable to saying: "It is not my duty to realize both p and not p." Here there is a sense in which it *can be* said that the above expression is logical, in that one could not be obligated to do two mutually exclusive or contrary acts at the same time, and in the same place. Again, however, this is not a strictly logical expression, as encountered in formal logic. On the contrary, it also cannot be construed as purely ethical, as would be the case with "It is my duty to protect human life." Herein is the difficulty in understanding the sense of the statements which Hallden deals with in all of *On the Logic of 'Better'*. For the expressions offered appear to be some species of hybrid of logical *cum* ethical statements, though without their being exclusively either one or the other. Moreover, Hallden assumes that in the context of his exposition one can adopt either an ethical intuitionism in his analysis, *a la* the early Moore, or an ethical naturalism, in the style of the empirical reductionism of Schlick. Yet even with these widely diverging ethical standpoints, what Hallden has to say about the logic of better could be accepted by a thoroughgoing decisionist, who bases choices on noncognitive action as in the works of Stevenson. At this point, one can perhaps excuse Hallden's laconic remarks on the actual meaning of "ethical validity" as an effort at not complicating the discussion by introducing questions concerning the foundations of various ethical theories. Nonetheless, there is a sufficient degree of obscurity on the point of how one is to construe "intuitively ethical validity," and "neutral comparisons," which get in the way of accepting the kind of thing Hallden wants to say. Surely, it is very much of a serious issue as to whether there can be a logic of the better and the same which cuts across all different kinds of ethical theories, as Hallden proposes. This is not to cast doubt in any direction of the issue. Rather, the point here is simply that it would have helped Hallden's cause enormously if he would have devoted some time to explaining how it would be possible to have a logic of the better and same, which is recognizably formal, and yet neutral with respect to any specific ethical and logical viewpoint.

---

[12] ibid., pp. 18-19

It *could* also be argued contrary to Hallden, assuming that one is dealing in the assertive mode, that the expression: 'pBq ≡ (pBq)B-(pBq)' clearly illustrates the law of excluded middle. Its acceptance seems based more upon logical considerations, rather than on how 'B' is found to be expressive of the *ethically convincing*. For one could say that when one endeavors to introduce in his analysis concepts which are "logical," as Hallden does with his free use of negation, propositional variable, implication, material equivalence, etc. then there must be some commitment to the system of logic which is being used to articulate the analysis. The acceptance of this expression as valid need not involve any of Hallden's allusions to "ethical validity," however the latter may be explained. At the very least one could argue that the expression: 'pBq ≡ (pBq)B-(pBq)' is illustrating the exclusive sense of disjunction, by means of 'B'. However, Hallden insists that there is more here than a logical relation.[13] For as was noted at the beginning, to say that p is better than q requires primarily that one also mean in a 'normative' sense that p is to be preferred to q.

Again, it is precisely the introduction of the "normative" sense of better, and therefore of preferring, which seems to be in need of clarification as to which, if any, ethical sense is being tauted. Hallden's claim that one can speak of the normative in a general sense without commitment to any ethical theory at all may be an issue of profound weakness. The first point which must be made is that if "normative" here is going to be taken in a totally neutral sense, i.e., it is going to be free from any and all philosophical presuppositions, then it will have to be intelligible in some "ordinary" sense. This suggests that the only way remaining in which the normative is to be understood must be the so-called "ordinary language" sense. That alternative would be totally rejected by Hallden, as shown in his opening comments. Secondly, it is not unacceptable to suppose that one may adopt an ethical viewpoint which is antithetic toward the importance of one's acting in a logically consistent manner. In essence, what is being suggested here is that there may well be a situation where one's ethical convictions may conflict with certain clearly logical criteria. Would Hallden be able to claim that his logic of the better is still applicable in this case as well? Rhetorically, "ought one," or "is it one's duty" always to act in a logically consistent manner? Thirdly, looking at the suggestion that the expression 'pBq ≡ (pBq)B-(pBq)' can be interpreted as: "It is not my duty to realize both p and not -p," one may well question whether there is an ethical sense of "duty" operating here at all. For where one has an understanding that the same

---

[13] ibid., pp. 18-19.

time and place are being supposed in the assertion of this sentence, then it is not a question of the logical or ethical untenability of what is to be done, but rather the *physical* impossibility of acting in two opposing ways. In summary, the point here is that if the expressions Hallden presents in his logic of the better are neither strictly logical nor solely ethical, but neutral, then the very intelligibility of such expressions is open to doubt, given that they are also divorced from any ordinary language signification.

Furthermore, to employ the "ordinary" mode of speaking about the "better" and "same," assuming that "ordinary discourse" is not another philosophical fiction, entails subscribing--in whatever provisional fashion--to the embedded preconceptions regarding these terms within the language itself. To proffer one's own views concerning the proper conceptualization of the meaning of these terms involves in some sense the illustration of how these terms, within ordinary discourse so-called, are somehow either adequately or inadequately perceived, or perhaps even totally misconceived. Thus, any suggestion of a "new" way at accurately conceiving of the "better" and the "same" can in a way be seen as either a clarification or a correction of an "ordinary" perception of such terms. Hence, the language one adopts or appeals to for whatever reason, has a great deal to do with the direction of the ultimate conception of the subject at hand. As was seen above, Hallden's very attack upon the ordinary sense of "better" and "same" was an effort at disowning this mode of speaking about these notions, and thus to intimate a clearer and more precise perception of these ideas. For this reason, it seems incongruous that after his explicit rejection of ordinary discourse as a vehicle for discovering an accurate conception of these terms, Hallden is found to appeal to an ordinary language context where he alludes to an "intuitive ethical validity," or to the "public's acceptance of the moral," or even to his vague reference of the normative in a *general* sense. These kinds of appeals to the "common" view constitute, in essence, a way of slipping back into "ordinary language" preconceptions at certain key points within his analysis. This will be the major point of criticism with regard to the conceptual preliminaries Hallden asks his reader to adopt. In summary, in his desire to achieve a puristic logic of the better, one which is totally sterilized from any nuance relating to ordinary discourse, he is seen to be returning to these very same modes of thought so as to explicate certain crucial aspects of his logic. It may be said that what is occurring in Hallden's analysis is something of the same phenomenon encountered in Ramsey's work, and in the presentation by Davidson, McKinsey and Suppes. Namely, the manifestation of the need to put a human face upon the highly abstract and, therefore, detached analyses that are generated.

Hallden appeals as well to the role of "intuition," as he lays the foundation for his calculus of the better. In this direction he proceeds to orient the reader, noting that the expression: pBp is "intuitively" self-contradictory. He notes as well that the intuitive understanding involved here is the direct result of what public moral acceptance has come to adopt, and there is thus no difficulty in explaining it. It is also in light of this public tolerance that one is to understand the idea of an "ethical conviction" of the truth of a proposition such as: 'pBq ≡ (pBq)B-(pBq)'.

Moreover, his allusion to an intuitive knowing of the truth of certain propositions, as for example the self-contradictory expression '(pBq)', comes very close to being an appeal to a general consensus for the meaning of "better." Again, it is difficult to explain the nature of this intuitive knowing other than in terms of a reasoned accumulative insight, which enables one to recognize the tenability or nontenability of some particular expression. However, one also discerns that an intuitive recognition of a self-contradiction still must involve some knowledge of formal logic, in that in this case its object, i.e., logical contradiction, is in essence being presupposed as the object of the intuitive recognition, so-called. Hence, is anything being added as substantive information where it is claimed that one has an intuitive recognition of a self-contradiction? The self-contradiction itself, as a logical concept, is the result of following certain pre-established definitional prerequisites. To add that one has also an intuitive awareness or understanding of the self-contradiction does nothing by way of justifying that one is in truth dealing with a self-contradiction. The point being made here is that expressions such as 'pBp' and 'pBq ≡ (pBq)B-(pBq)', if they are going to be said to have any logical character, do so in virtue of the system of formal logic, i.e., in a system of analytical truths based upon definition and not upon an allusion to intuition. Yet, one must be reminded again of the fact that Hallden insists that statements such as the above are *not strictly* logical.

In surveying Hallden's enterprise at this juncture, it appears that not only is there a question as to the soundness of the distinction between logical and ethical validity, but the apparently key idea of an intuitive understanding as somehow permeating his analysis and lending certainty to his insights is suspect. Furthermore, in insisting upon an affinity between an intuitive recognition of validity and public recognition, it can be argued that Hallden is once again, through a different channel, re-introducing into his analysis all the variability of meaning which ordinary discourse imposes, and which he explicitly desires to avoid.

Hallden himself senses the difficulty in his own position where he says earlier in his work that there does not seem to be a definite way of differentiating between the total effects of "better" and "same," and the so-called neutrality of both these terms.[14] In essence, he is recognizing that one cannot divorce the empirical elements from a purely formal framework when presenting a logic of preference. However, apart from this shortcoming in a central aspect of his logic, Hallden proposes that his approach will somehow escape the empirical constraints he finds so disconcerting. In a sense he is saying that a logic of the better seems always to be limited in that it must reflect the totality of possible effects which saying one thing is better than another involves. Yet in granting this much, the same question arises again, namely. how a logic of preference can ever be *formal*, i.e., visualized with the help of symbolic logic, which is the approach Hallden has adopted?

The conceptual foundation Hallden wants to build upon is one where there is an "intuitive justification" for saying that p is better than q. This is taken as a logical insight which reflects an inner automatic assessment of the value of one state of affairs in contrast to another. Yet, involved with this in some way not explained is the collective understanding of a "moral necessity" which pertains to the sense of saying that p is better than q. The morally necessary here is left open as a concept, i.e., it is applicable to literally any moral outlook. In other works, e.g., *Emotive Propositions*, he makes reference to G. E. Moore and the "common sense" meaning of ethical terms. However, one is not certain whether the sense of the morally good or true Moore alludes to would do much to illuminate Hallden's conception of the morally necessary, since it is intentionally left undefined; which is to be distinguished from discussing the issue of the tenability of even operating in this manner.

In view of the above difficulty, one may doubt the fruitfulness of pursuing any further study of Hallden's work. Yet, to simply dismiss Hallden's contribution on the overall point of its failing to provide a logic of the better which is readily comprehensible because of its not referring to concrete cases, is to lose sight of how his work constitutes an innovative departure from what had occurred in the past in this area. Moreover, it is also to fail to appreciate Hallden's superb mastery of logical technique.

The unfolding of what Hallden has to say illustrates his commitment to the belief that a purely formal analysis, coupled with intuition, can help one arrive at a logical topography, so to speak, of any and all discourse involving

---

[14] ibid., pp. 15-16.

the better and the same. The tools Hallden employs are not derived from economics, contrary to what was seen in the case of Davidson's attempt a few years earlier. Rather, Hallden tries to remain wholly within the domain of formal logic, attempting to engineer a logical structure for reasoning regarding the better. The intricacy of the edifice he creates is intriguing in its complexity. Though, as in the past, the problem ultimately comes down to one of the relation between form and content, Hallden charts a new direction in what he is proposing. In this he deserves recognition as a pioneer in the development of a logic of preference.

II. *Theory A*

Hallden's first theory of preference, or Theory A, is designed to show how on an elementary level a formal logic of preference encounters limitations with respect to certain distributive rules.

Hallden begins by stating that the 'B' relation is asymmetrical and transitive, and that relation 'S' is reflexive, symmetric and transitive. Within this context he asserts that the expression: "... the better the presence of something, the worse the absence of it ..." is defensibly valid.[15] Hence, he sets down the first postulate of his logic of preference as follows:

I. $pBq \supset -qB-p$

What makes (I) appear peculiar according to Hallden is that it is expressed in uncommon negative terms. Its negative modes of expression makes (I) look counter-intuitive, whereas in essence it is a valid expression of the implication that state of affairs p is better than that of q.[16]

From (I) Hallden derives the following axioms,

A7. $pBq \equiv (p.-q)B(q.-p)$
A8. $pSq \equiv (p.-q)S(q.-p)$

It is to be noted that (I) above is presented in terms of what would, be the preferred state of affairs, given two alternatives. This is to say that if p is going to Paris, and q is going to Algiers, Hallden observes that one considers whether it is preferable to go to Paris *without* going to Algiers. The key point here is that state of affairs p is considered in the absence of q, i.e., in terms of p.-q. Consequently, in considering the "better than" relation, one is invariably introducing the negative of one of two possibilities, though one is not consciously aware of the introduction of the negative. Hallden relegates the peculiarity of expressions like (I) to the "psychological expectancy" inherent in ordinary conversation, *where negative facts* (e.g., -qB-p) in the sense of (I) and A7, as

---

[15] ibid., pp. 25-27.
[16] ibid., p. 27.

something immediately evident, and not usually noticed as operating in the background of the expression.[17]

Perhaps the one point of universal contention in various commentaries on Hallden's work is the soundness of his view that evaluative expressions covertly imply an aversion on the part of some agent toward a negative state of affairs. This was claimed to be the case in (I) pBq ⊃ -qB-p above, as well as in A7. It is a major contention throughout Hallden's presentation that implicit in all formal expressions concerning the better there is an allusion to certain "negative facts," which the agent in some indirect manner wants to avoid. The justification for this belief emerges from a fundamental ethical intuition which governs his views on the nature of the logic he is developing. This comes out in a striking manner in his defense against attacks by Castaneda, Jeffrey, Chisholm, Sosa and others.[18]

Recognizing the intuition in question involves taking into account what Hallden terms in later publications the principle of "Ethical Concretism." The latter presupposes that any (ethical) proposition which asserts the existence of something, and in no way denies the concrete existence of anything, will be called a "concrete existence propositions." Thus, given the two propositions p and q, both will be taken as equivalent to each other if: (1) if p ⊃ r, and r is a concrete existence proposition, then q ⊃ r, and (2) if q ⊃ r, and r is a concrete existence proposition, then p ⊃ r. Briefly, the Principle of Ethical Concretism states that where p and q are equivalent with respect to r, then p is not better than q. This principle, however, is seen to contradict "common sense" ethics, and especially postulate (I): pBq ⊃ -qB-p. The examples Hallden cites to prove this point are taken from the standpoint of a general hedonist viewpoint:[19]

"(9) From Jan.1, 1970, onwards, no one will feel any pain.

(10) From Jan. 1, 1970 onwards, no one will feel any pleasure."

The point here is that both (9) and (10) make no explicit statement about the actual existence of pleasure or pain. Thus, in accordance with the principle directly under consideration (9) is no better than (10). However, from the viewpoint of ethical hedonism, (9) must be better than (10). For Hallden, this supports the claim that it is an intuitive certainty which is reflected in I. Consequently, to claim that p is better than q, must entail in ethical discourse that p and not -q is better than q and not -p. Within the purview of ethical

---

[17] ibid., p. 29.

[18] ibid., *Logik Rätt och Moral, Filosofiska Studier tillagnade, Manfred Moritz,* 1969.

[19] ibid., pp. 72-73.

discourse in general, there is an intuition of certainty operating which is reflected in the kind of relation which (I) is expressing. This is a further justification of the thesis that the logic of the better is reflective of certain intuitions found in the examination of ethical propositions generally. In contrast to the Principle of Ethical Concretism, Hallden wishes to point out that his postulates and axioms for a logic of preference do not assert anything concerning an existential state of affairs, and thus earn their credibility strictly from an intuitive perspective.[20]

Hallden also defends the truth of Postulate (I) from a number of attacks by commentators. First is the evaluation of Castaneda's criticism of (I), with respect to the latter's claim that it leads to a self-contradiction. Castaneda denotes three states of affairs, p: saving $1,000, q: saving $950, and r: saving $900. Quite simply, Castaneda observes that (1) q.r.B.p, and (2) pBq. Consequently, it can be said that qr.B.q, according to the transitivity of B, together with A2 and B2. By the application of Postulate (I) one discovers that qr.-q:B:-(qr).q, which for Castaneda is said to be the same as q-q.B.q-r. Castaneda does not explain why the latter is derived from an instantiation into (I); however he does note how the latter is allegedly self-contradictory in that it results in saying that one's both saving and not saving $950 is better than saving $950 and not saving $900. Given that 'ought' implies 'can' for Castaneda, he just cannot allow for the possibility of acting in two contradictory ways, namely; q-q and q-r.[21]

Justifiably, Hallden questions the tenability of Castaneda's conclusion on the last point. For it does not seem that q-q.B.q-r can be logically inferred from an instantiation into (I). Moreover, Hallden notes that the three states of affairs Castaneda mentions are not properly construed by him. This tends to mar the incisiveness of Castaneda's review of Hallden's book. Specifically, Hallden points to the three acts which the above state of affairs involve better presented as:

"(1) Both p and q are saved.

(2) Only p is saved.

(3) Only q is saved."

Instantiation in Postulate (I) results in saying: "(1) and not (3) is better than (3) and not (1)." However, "(1) and not (3)" does not express a conjunction

---

[20] ibid., p. 73.
[21] Castaneda, Hector Neri, "On the Logic of the 'Better'", review, *Philosophy and Phenonmenological Research*, 1958, Vol. 19, p. 266.

between two contrary acts, since for Hallden (1) and (3) express "different" or "distinct" acts. This is apparent from the fact that the word "only" is introduced to indicate that in (2) *only* the act of saving $1,000 is being considered, and in (3) *only* the act of saving $950 is involved. Thus, one must not relate the acts described in (1), (2) and (3) in terms of their accumulative effects, but rather each act should be seen in terms of its own specific value of betterness. In essence, one could say that Castaneda is assuming an ethical concretism in his criticism of the tenability of Postulate (I) which Hallden implicitly rejects. For Hallden, Postulate (I) is meant to reflect a betterness-relation which concerns the uniquely intrinsic value of states of affairs involved in moral discourse.[22]

Hallden's rebuttal of Chisholm and Sosa's criticism follows along the lines seen above. The latter submit the following as a necessary contradiction resulting from the application of A7: pBq p.-qBq.-p: "... Smith and Smith's wife being happy while Smith is not happy: (p.q).-p is better than that state of affairs consisting of Smith being happy and it being false that both he and his wife are happy: p.-(p.q.)." or (p.q.).-pBp.-(p.q). Hallden notes that in this case two possible interpretations come into play when considering the sense of (p.q.).-p. These are manifested as:[23]

"(1) It is better that both Smith and his wife are happy than that only Smith is happy.

"(2) It is better that it is certain that both Smith and his wife are happy than that Smith is certainly happy and his wife possibly unhappy."

From Hallden's perspective, the examples Chisholm and Sosa present involve the comparison of logically incompatible things, and are thus not true to the kind of relation expressed by A7. For example, in (1) above, one is comparing the case of both Smith and his wife *being presently* happy, with only Smith *being presently* happy. Thus, the example itself is inconsistent since it involves an untenable (ethical) concretism.[24]

Perhaps Richard Jeffrey's objections to A7, and Hallden's subsequent rebuttal, will suffice as a conclusive illustration of the way in which the latter insists upon the plausibility of the concept of "negative facts" as essential to a

---

[22] Hallden, Soren, *Logik Rätt och Moral*, pp. 76-77.
[23] ibid., pp. 76-77.
[24] ibid., p. 77.

logic of preference. Jeffrey presents the interesting example of the Roman matron who prefers death to dishonor. Consistent with the application of axiom A7 one would expect to have the following:

(1) A: The death of the matron by next week is better than B: the matron being dishonored this week, being equivalent to (2) -B: the matron not being dishonored this week is better than -A: The matron not being dead by next week. Though one has here what appears to be a straightforward application of A7, what results in the latter component of the equivalence is a statement which evidently does not follow from saying that the matron prefers death to dishonor. Consequently, Jeffrey uses this example as a means of showing that the appropriateness of A7 for a logic of preference is suspect.[25]

Hallden, however, perceives that the matter has not been correctly stated by Jeffrey. For Hallden does not see that the Roman matron prefers -B to -A, but rather that she prefers death without dishonor or A & -B to dishonor with death or A & B. As in the case of Castaneda's example with the saving of money in contrast to not saving anything at all, and in the case of Chisholm and Sosa's situation where both Smith and his wife being happy excludes the possibility of *only* Smith being happy, so also the matron's preference of death and honor precludes the possibility of her living and being dishonored. In other words, Hallden is saying that it is of the nature of expressions of betterness or of preference situations generally that they "analytically" exclude that specific state of affairs which is opposed to the point of the preference. This is what Hallden conceives as a "negative fact."[26]

However, apart from Hallden's assurance that one need not be worried about the meanings of Axiom A7 and Postulate I, there are some very serious questions which emerge in the unfolding of Theory A. First, it is important to observe that Hallden refers to 'p' and 'q' as variables for "possibilities" or what appear to be possible world states. He also speaks in terms of the "presence" of p entailing the "absence" of q. Earlier, when attempting to set forth what he takes to be the positive sense of "S", he says "... It is presupposed that equality in value between two possibilities means something more than merely the *absence* of a definite order of value (my italics)."[27] Allusions to the presence or absence of some possible-world state, as well as references to the compara

---

[25] ibid., p. 77.
[26] ibid., p. 77.
[27] Hallden, Soren, *On the Logic of 'Better'*, p. 26.

bility of values, tend to burden the notion of "purely" formal implication, intended in Hallden's sense of entailment. Somehow his allusion to *actual* conditions of weighing the value of possible outcomes in given situations interdicts his stated aim of deriving a logic of the better which is purely formal, and totally independent of ordinary discourse. The question here is what he may mean by the "presence" or "absence" of a state of affairs. It really could not mean any "actual" occurrence in space and time, which will surely throw off the supposed hypothetical character of expressions such as pBq. This is to say that the expression p is better than q must at the very least mean that generally p is preferred to or better than possibility q. Thus, pBq cannot allude to an actual spatiotemporal occurrence while it is claimed that Theory A is purely formal. On the other hand, he is seen to claim that the betterness of p over q, or to the equality of p with q, is determined by the value of p and of q. However, it was also seen in his introductory remarks that the assignment of values is dependent neither upon intrinsic considerations, nor upon considerations dealing with the "total range of effects" for a possible world state. Somehow the attribution of value to p and q must reflect a neutrality from context and intrinsity. Moreover, this very insistence upon neutrality contradicts his attempt to explain the so-called negative aspect of Postulate I. discussed above, which is presented in terms of the negative *consequence* of p and q.

Thus far, Hallden trades heavily on revealing the intuitive validity of his postulates, as for example in (I) above. Yet one must press the point of what he thinks this "intuitively evident" is. Somehow it appears to be a kind of insight anyone can have once the correct empirical example has been brought forth. Again, however, the very allusion to such an example apparently negates the supposed neutrality of his logic of preference, so-called. At least there seems to be reneging on the claim that the total consequences of expressions involving preference are not reflected in the logic itself. Though earlier Hallden grants that there may never be a separation between the so-called neutral stance for attaching truth value to p and q, and the stance which requires considering the role of the total effect of p or q on the context in which they are considered.[28] At this juncture, it is seen that this claim works against the idea that there can be a purely formal logic of preference as envisioned by Hallden. The very possibility of introducing the empirical element so as to justify the allocation of value to a possible-world state, and the admission that this is an unavoidable consequence in spite of the adoption of a neutral stance, vitiates the formal character

---

[28] ibid., p. 16.

of the logic he is proposing. For the contextual element, with its concomitant extension, makes any attempt to introduce entailment between proposition a dubious enterprise at best. Consequently, not only is the form of such a logic problematic, but the entire notion of intuitive evidence, upon which Hallden bases so much as far as the justification of the axioms are concerned, is an obscure foundation upon which to proceed. Surely, he could not be taking the intuitively evident to mean the empirically justified, for then intuitive validity would be based upon a kind of inductive generalization, which would be foreign to the kind of formal character he desires his logic to have.

In Hallden's own terms, Theory A is seen as "logically weak." Basically, in criticizing A he makes somewhat the sort of observation Kenneth Arrow sees where the latter states that it does not necessarily follow that, because a is preferred to b, and b is preferred to c, that a must be (necessarily) preferred to c. Arrow demonstrates how everyday situations simply need not or actually do not obey the law of transitivity. In a parallel line, though not specifically referring to Arrow's recognition of the above point, Hallden sees the limitations inherent in Theory A as far as distributive relations are concerned. Hallden argues that the fact that two things (individually) are preferred to a third, does not imply that the former two both in conjunction are preferred to the third. Thus, the following formulas of A are not in line with common sense "ethical" evaluations.[29]

H1. $pB(q.r) \supset (pBq.pBr)$
H2. $(pBq.pBr) \supset pB(q.r)$
H3. $(p.q)Br \supset (pBr.qBr)$
H4. $(pBr.qBr) \supset (p.q)Br$

In all of the above, the implicational relation simply cannot be said to express how the evaluations of state of affairs *individually* justify that some *conjoined* permutation of these states of affairs is preferred.

In this vein Hallden goes on to claim that "It can never be proved by arguments which are purely logical that one thing is better than another ..."[30] The above, together with the previous allusion to an incompatibility with common sense evaluations, suggests again that he is unhappy with the way in which Theory A falls short of expressing the variety in occurrences involving value. As was seen in earlier attempts at formulating a logic of preference, there is here as well the now familiar concern with the nature of "fit" between the

---

[29] ibid., pp. 39-40.
[30] ibid., p. 42.

logic and the elusive context of everyday occurrences as alluded to in natural discourse. Somehow, the logic always seems to fall short in its capacity for reflecting the variety of the "ordinary" circumstances. Relations presented in terms of propositional calculus cannot be made to represent precisely the nuance of implications which is found in discourse concerning the better.

The same concern with the limitation of A is found to permeate Hallden's discussion on the Comparability Hypothesis. The latter, expressed as: H10. pBq v pSq v qBp is designed to express *all possible* relations between p and q within Theory A. Hallden is ambivalent as to the tenability of H10., largely because it does not seem to translate into something which makes sense in an everyday real world situation. H10's ethical and logical validity cannot be supported by appealing to common sense. For one can envision the case within an ordinary context where p and q are unrelated to each other given a specific value scheme. Thus, the indifference of p to q, and conversely, simply undermines the universal applicability of H10. as a useful tool in Theory A.[31]

The same difficulty in relating the context Hallden is attempting to formulate with purely logical expressions appears where he sets up matrices for proving hypotheses in Theory A. In proceeding, Halldan stipulates that 'S', the relation of sameness, is to be interpreted in the propositional calculus as 'p v-p'. He argues that with the foregoing assumption one can demonstrate the consistency of formulas H1. to H10. above.[32] However, what does it mean to say that 'pSp' can in some sense be equivalent in meaning to the expression 'p v-p'? Often in his text a similarity is noted between an expression dealing with something being better or the same and some expression in formal logic, but there is very little by way of explanation as to what constitutes this similarity. In propositional logic, connectives are given a very precise definition relative to their truth functions. However, the better or the same as evaluative terms seem fraught with references to "intuitive validity" specifically appropriate for discussions involving ethics. In a sense, one seems to be constantly in the position of comparing apples to oranges when endeavoring to read a "logic" in Hallden's formalization of preference, at least as it is presented in Theory A.

Theory B

The stated purpose of the second theory Hallden presents is to concentrate upon the justification for saying that B-formulas are logically valid. Accordingly, he holds that Theory B will be more accessible to metalogical consid-

---

[31] ibid., pp. 45-47.
[32] ibid., pp. 54-55.

erations. Furthermore, Theory B will include Theory A, together with the Hypothesis of Comparability. In addition, B contains variables 'V' and 'T' respectively, ranging over variables and terms with suitable subscripts. The usual rules of modus ponens, substitution, and replacement are assumed in B. The foregoing together with the following postulates constitute the core of the Theory:[33]

> B1. pBq ⊃ -(qBp).
> B2. pBq.qBr. ⊃ pBr.
> B3. pSq
> B4. pSq ⊃ qSp.
> B5. pBq.qSr. ⊃ pBr.
> B6. pBq ⊃ -qB-p.
> B7. pBq v pSq v qBp.

One of the most intriguing proofs Hallden offers in his presentation of Theory B deals with his devising of a decision procedure for determining the validity of any given formula of B. The latter introduces mappings, univocal expansions, and multivocal expansions.

First, no term in a map, which is any atomic formula of B, contains more than one negation. Also, conjunctions of atomic formulas exhibiting this characteristic are maps. "No other expressions are maps."[34]

Considering for example the map 'pS-p . qS-q . pBq' one sees that the expression is explicit with respect to indicating the nature of the relation between p and -p, q and -q, and between p and q. However, it is deficient in expressing anything concerning the nature of the relation between -p and -q, p and -q, as well as explicating the relation between -p and q. Thus, the map is incomplete until it can be shown that it can undergo a process of expansion, i.e., *univocal expansion*, so that the above indicated unexpressed relations can become explicit.

The expansion of the map will be presented in its entirety so that the reader can appreciate the ingenious character of Hallden's proof, and thus perhaps go on to a closer study of this beautiful work. For clarity, the rules Hallden employs for the various transformations will be provided for each step of his proof.

---

[33] ibid., pp. 59-61.
[34] ibid., p. 65.

Given map:

pS-p . qS-q . pBq	*Justification*:
1. pS-p . pS-q . pBq . -qB-p.	(b.19, pBq . pBq . -qB-p)
2. pS-p . -pSp . qS-q .-qSq . pBq . -qB-p.	(b.14, pSq.pSq . qSp)
3. pS-p . -pSp . qS-q . -qSq . pBq . -qB-p . -qB-p . -qBp.	(b.16, pBq . qSr pBq . qSr . pBr.)
4. pS-p . -pSp . qS-q . -qSq . pBq . -qB-p . -qBp . qB-p.	(b.17, pSq . qBr pSq . qBr . pSr)

With step 1, one has an explicit statement as to the relation between -p and -q. In step 3, one has the expression of the relation between -p and -q. Finally, in step 4, one has the explicit rendering of the relation between -p and q.[35]

The map considered above can be expanded by further applying Hallden's rule b.19, which states: pBq . pBq . -qB-p. Thus, the new expansion goes on to say:

5. pS-p . -pSp . qS-q . -qSq . pBq . -qB-p. -qBp . qB-p. pB-q. Step 5, however, can be further expanded by the application of b.15, which states: pBq . qBr .≡. pBq . qBr . pBr. The resulting expansion involves the expression of an impossible state of affairs, namely that of -qB-q. For the latter must be false according to Hallden's rule b.1, which claims that -(pBp).[36]

Thus, it is seen how the continued univocal expansion of the map to its final completion can produce a proof for the map's own negation. Hallden notes that the self-contradiction which occurs is a syntactical concept, since a term cannot be said to be better than itself, e.g., *TBT*. What allows one to speak of the expansion of a map in abstraction is, in essence, this rendering in syntactical terms of the notion of self-contradiction. From the above, Hallden also arrives at four other concepts which are of importance for carrying out additional proofs in Theory B. First, the notion of a *forbidden* map is one of a map like the one explained above where a self-contradiction occurs. On the other hand, a *permitted* map is one where its expansion does not result in its manifesting a self-contradiction. A map is said to be *complete* if and only if given two distinct terms T and T_1 occurring in it (say map P), then at least one of the following formulas is expressed in P: *TBT₁*, *TST₁*, *T₁ST*, and *T₁BT*. Where none of the forementioned formulas occur in a map, then that map is said to be

---

[35] ibid., p. 65.
[36] ibid., pp. 65-66.

incomplete. It is to be observed that the concepts of a permitted map, forbidden map, complete map, and incomplete map are all syntactical.[37]

Some trivial results of the above discussion of the univocal expansion of a map may be noted at this point. First, it is readily seen that if formula Q is seen to be forbidden, then Q is not a theorem of B, and -Q is a theorem of B. Also, if Q is the result of the expansion of P, and Q is forbidden, then -P is a theorem of B. Moreover, one notices the application of the Hypothesis of Comparability, so-called, in the univocal expansion of the above map. This is to say that the determination of the completeness of an expanded map depends upon the appearance of permutations of the better or the same precisely in the way afforded by the forementioned hypothesis.[38] Interestingly, at this point Hallden does not voice concern on the matter of whether or not relations noted by the expansion cover the range of possibilities involving indifference. Since in the present discussion completeness is presented as a syntactical relation involving the possible concatenation of *terms* relative to B and S, the issue is not as sensitive as in the discussion of the hypothesis itself in Theory A, where the object of concern was the range of possible relations between variables.

A problem of importance for Hallden is that of devising a decision procedure for determining whether a map is a theorem of Theory B. He introduces a procedure for solving this problem which he terms "multivocal expansion." Without entering into the complexities of the particular proof Hallden alludes to, it is perhaps more to the point to briefly survey the rules which govern this kind of operation.

The first step requires assuming that the atomic formula P is univocally expanded and that: (a) either T or T' occur in P, and (b) either T or T'_1 occur in P. Furthermore, P is assumed to be incomplete, relative to the occurrence of the formulas TBT_1, TST_1, T_1ST, *and* T_1BT in B. Consequently, P can be expressed as a conjunction of the forementioned formulas with P. It is also supposed that P is not univocally expanded, and that Q is the end product of P. Thus, P is replaceable by Q. Generally, multivocal expansion comes to an end either when one of the formulas of a group resulting from an expansion is permitted, univocally expanded, and complete, or when all of the formulas of such a group are univocally expanded and complete. Trivially, what assures the progress of multivocal expansion is the logical relation of conjunction.[39]

---

[37] ibid., pp. 66-67.
[38] ibid., p. 68.
[39] ibid., pp. 72-74.

Hallden terminates this aspect of his discussion of the mapping of atomic formulas by noting how it can be said that if the multivocal expansion of P is carried through, then P is a theorem of B, and -P is not a theorem of B. On the other hand, where the multivocal expansion of P yields a formula or group of formulas which are forbidden, then P is not a theorem of B, though -P is a theorem of B, etc.

Having prepared the groundwork for the operation of multivocal expansion, Hallden proceeds on to a more ambitious proof, namely that of showing how -P is not a theorem in theory B, where it is given that P is permitted, univocally expanded, and complete. Preparation for this proof is intriguing since it involves the introduction of a matrix M, where the values of T and F are considered from three distinct modes: i.e., the Positive, the Neutral, and the Negative. Consequently, Hallden goes to great lengths to show that an isometry exists between operations performed within the propositional calculus, PC, and the satisfaction of the matrix. The result is that he derives the crucial metatheorem that if P is a PC-theorem, then it is satisfied by M. Furthermore, it follows that if P is a theorem in theory B, then P is satisfied by M. From these steps, Hallden takes the additional step of ordering the terms of B according to the numbers 1, 2, 3, ... Hence, one sees Hallden alluding to the rigor of mathematical order so as to suggest a more precise relationship between the terms of B.[40]

Hallden's introduction of mathematics as a means of making more precise the relations between the terms of B is different from later uses of mathematics in developing logics of preference, especially in Rescher's attempt in 1968. Here the mathematical ordering is given as a way of tightening the syntactical relations between the terms of B, whereas Rescher uses mathematics as a way of handling the evaluation of states of affairs, which come to constitute the options of a preference. Consequently, Rescher's innovation will be to involve mathematics more into the semantics of preference, whereas Hallden restricts the use of mathematics within the confines of the clarification of syntax in a formal theory of preference.

Upon showing the correlation between the terms of map P with a simple mathematical progression, Hallden proceeds to demonstrate a correlation between the terms of P and the indices of the elements belonging to matrix M, i.e., the values of M. His strategy is to define a numerical correlation with matrix M, expressed by '$M(T)$', as a function of two numbers, namely as: (1) the

---

[40] ibid., p. 75.

correlation of term T with a number, expressed as the function: $N(T)$, and (2) with the highest of all numbers, i.e., 'n', correlated with the terms of map P resulting from the first correlation: $N(T)$.[41] With the preceding fully explained in terms of lemmas illustrating the various consequences of correlating terms with numerical progressions, and the assignment of numerical indices to the value-elements of matrix M, Hallden holds that he is in a position to show that - P is not a theorem of B, where P is permitted, univocally expanded, and complete. Very simply, he demonstrates how the indices for the value T-elements of P will be determined according to the following rule: "if $M(V) = i$, then T_i is assigned to V," where 'V' stands for any variable of B. In a series of four cases where TBT is as representing any of the following four types of formulas: VBV_1, $-VBV$, $VB-V_1$, and $-VB-V_1$, Hallden shows through an elaborate sequence of steps that the value of TBT_i must always be T_i, given P. Hence, it follows that $-P$ will have a value of F_{-1} which means that $-P$ cannot be satisfied by matrix M.[42]

Though an exhaustive presentation of Hallden's intricate mathematical proof would be basically a repetition of his own analysis, for the purpose at hand it can be said that enough has been demonstrated to illustrate the direction of his thinking, and how he is aiming at a purely formal theory of the better. The remainder of his investigation pursues the basically formal aspects of his analysis to show the semantic completeness of B. However, to recall Castaneda's laconic comment in his review of Hallden's work, there is very little by way of "philosophical value" in the latter's formalization of preference. Underscoring Castaneda's brief remark, one can note the pervasive dominance of the manipulation of the symbol on the syntactical level. Though in doing this Hallden displays an admirable sense of handling logical formalization, it is still at the expense of losing the "relevance" of his logic to specific forms of human experience. This is to say that the human element, with all of its interesting facets dealing with immediate concerns and the measuring of possible consequences, seems lost in the emphasis upon the "logical rigor" of his analysis. This is not to say that his formal analysis is incorrect in substance, rather his approach seriously begs the question of the "use" of this study in elucidating the concept of preference as employed in everyday usage. Granted that there is a measure of truth in Hallden's observation that common usage is fraught with ambiguity and thus defies precise formalization, what emerges at the

---

[41] ibid., p. 79.
[42] ibid., pp. 80-81.

conclusion of the present treatise is a highly sterilized system, shedding little light upon the "logic" of discourse involving preferences.

Again, within the context of work appearing prior to 1958, Hallden's contribution can be considered as an attempt to by-pass the issue involving the nature of the relation between theory and the object of one's analysis, namely preference as ordinarily understood in everyday circumstances.

In reviewing the importance of the area of study he helped pioneer, it is best to allow Hallden to speak for himself through his most recent correspondence with this author:

> "... When I wrote *On the Logic of Better*" I felt uncertain about the philosophical importance of the field. My personal interest was focused on the methodological aspect — I wanted to find out about the relevance of the logicians craft for the intuitive analysis of fundamental philosophical concepts. Afterwards, in (the) beginning of the sixties I became interested in the pragmatic foundations of science of Reichenbach and Wisdom, and started work in what I called "strategy of science". I now firmly believe that the sure-thing principle of Leonard Savage constitutes one of the foundation stones of human knowledge. And this principle is closely connected with my own "if pBq, then (-q)B(-p)". This has of course changed my view of the relevance of the field."

Chapter 6.

The Many Modal Interpretations of Prohairetic Logic: Aqvist, Chisholm, Sosa and Hansson

Along with the two towering achievements by Soren Hallden and Georg Henrik von Wright are some less in-depth though equally interesting attempts by philosophers, who while affiliated with Hallden and/or von Wright in their use of an intuitionistic approach, see the issue of evolving a logic of preference differently from what is found in the above two cases. This group of thinkers, composed mainly by Lennart Aqvist, Roderick Chisholm, Ernest Sosa, and Bengt Hansson, represent *in part* a historical connection between Hallden and von Wright. Collectively their work constitutes a strong statement as to the influence which Hallden's treatise had on his contemporaries. In each of the studies to be considered below, one invariably finds some reference to Hallden's incisive work, as well as the expression of a need to seek a different direction, by way of improving upon what the former had achieved. Individually, however, these writings exhibit a partial historical continuation, in the sense that though Aqvist writes between the time of Hallden's publication and before the appearance of von Wright's treatise, Chisholm, Sosa and Hansson present their views subsequent to von Wright's monongraph in 1963, and are thus not entirely in sequence before von Wright. Their work, however, is influenced more predominantly by Hallden than by von Wright, and thus a discussion of what they have to say deserves attention here rather than further on.

In the sections which follow it will be shown how each of these three major efforts, composing the material bridging the gap between Hallden and von Wright, illustrate an innovative move in the historical development of this area of philosophical analysis. For example, Aqvist fully explores the possibility of using modal logic to express the formalization of preferences. This is an as yet untried approach, and one which relies perhaps too heavily upon certain basic and yet not easily defended presuppositions of deontic logic itself. Chisholm and Sosa see the importance of articulating the logic of preference within the context of moral discourse involving the hedonically good. This is again a new approach to the problem, an innovation which is furthered by their keen interest in coming to grips with a correct characterization of the "objects" with which a logic of preference deals. Their profound understanding of Brentano and Meinong lends depth to their analysis of the ontology of preferential expressions, more so than in any other effort which is part of the philosophical literature for this area of inquiry. Finally, Benght Hansson endeavors to develop a logic of preference which is totally context-free, one which reflects a flexibility so unique, it is believed able to exhibit the formal properties of preferences in any context whatsoever. His aim is to create a logic of preference which is able to represent the fluidity preference has in "ordinary talk." In this, one sees a further change in direction. This is to say that Hansson sees no reason why a logic of preference should be expressive *only* of how preferences are structured in moral discourse, as is held to be the case in the contribution by Chisholm and Sosa.

In all of this, there is a single thread which connects these three different approaches to the problem. This is the interest in the "modality" of preference. One sees this in Aqvist's manner of attempting a deontic logic of preference, where preferential relations are expressed as reducible to the obligatory or the forbidden. Similarly, with Chisholm and Sosa the emphasis is upon the intrinsicity of preference, with the proviso that the preferred is that *possible* state of affairs which is the state of the highest pleasure. In Hansson, the intent is more sharply focused upon the extraordinary flexibility of the notion of preference in ordinary discourse, and how only a modal logic of preference can capture its subtlety. Thus the multi-facetness of the modal, as the obligatory, the possible, or the good — considered in the broad and not necessarily moral sense encountered in ordinary discourse, is the unifying conceptual framework in which Aqvist, Chisholm, Sosa and Hansson present their views.

Finally, it is interesting to observe how a subtle shift in emphasis is beginning to manifest itself, by way of a gradual awakening to the realization of the need to attend to the function of ordinary discourse when articulating a logic of preference. Such a change is suggested indirectly by Aqvist's interest in the obligatory, etc. as a basis for a deontic logic of preference. This it is somewhat escalated in Chisholm and Sosa's attention to the intrinsic sense of preference in "moral discourse." Ultimately, it is given prominance by Hansson's accent upon the formalization of preference in ordinary discourse. This movement will, of course, accelerate as new efforts are introduced. From here onwards, the role of ordinary discourse as an object of inquiry for a logic of preference can never be neglected, as was the case in the past.

Section I. *Aqvist's Logic of the Deontically Better*

Aqvist's departure from Hallden lies in the former's insistence that the deontic logic of better should be articulated not on the basis of value concepts such as the morally good, bad and indifferent, but rather that it should be based upon deontological concepts reflecting the "obligatory," the "forbidden," and the deontically "indifferent."[1] It is very important to keep clear at the outset that Aqvist's foremost interest is in presenting a deontic logic, which is founded upon the notions of the better, same or worse. Conceptually, this stands in contrast to the work of Hallden, who though in highly critical terms saw his logic of better as being free of the ambiguities which evaluative terms invariably incur in everyday discourse involving morals, nonetheless held that his logic reflected a "purer" view of the evaluative aspect of the notion of better. Aqvist, on the other hand, endeavors to present a deontic logic of preference, based upon the idea of the "propriety" of acting in certain clearly defined situations.

Aqvist's position is even more radically opposed to von Wright's, since the latter, within the same year in which the former proposed his deontic logic of preference, questioned the viability of any deontological approach toward explaining the notion of preference. Both in his classic treatise, *The Logic of Preference*, and in his more general study, *The Varieties of Goodness*, von Wright argues that deontological notions cannot be easily disassociated from evaluative ones, and thus deontology, as a specific area of philosophical inquiry, is beset with inescapable ambiguity. Though not referring to Aqvist directly, von Wright argues in *The Varieties of Goodness*, that imperatives, deontic sentences, and *anagastic* sentences are all such that in ordinary discourse there is often no clear point of distinction between the sense of saying "You ought to close the window." "Close the window!" and "You must close the window."[2] On account of the inherent obscurity in the very notion of the obligatory, von Wright will be found to proceed on a different tract in developing his own logic of preference; one which though intuitionistic like Aqvist's, seeks to accommodate more varied components of preference expressions, e.g., events, acts, states of affairs, etc.

---

[1] Aqvist, Lennart., "Deontic Logic Based on a Logic of Better", *Acta Philosophica Fennica*, fase, XVI, (Helsinki, 1963), p. 285.

[2] Von Wright, Georg Henrik., *The Varieties of Goodness*, Routledge & Kegan Paul, 1963, London, pp. 157-158.

By way of clarifying the perspective from which Aqvist is operating, it is instructive to consider a brief comparison between him and von Wright. At least in the preliminaries, it is plain to see that whereas von Wright concentrates upon a formal characterization of preference in itself, that is, as a notion which has its own unique identity, and therefore its own logic, Aqvist treats the notion of preference in a derivative sense, in that he sees it as definable in terms of deontological concepts. Thus in a sense it may be said that for Aqvist the logic of preference is an extension or application of deontological analysis. Both in the essay now under review, and in an earlier essay entitled, "A Binary Primitive in Deontic Logic," Aqvist cautiously describes his task as that of articulating a deontic logic on the basis of the binary relational term 'B', which *might* be identified as related to the concept of "betterness."[3] Here it seems reasonable to conclude that his admitted candor in suggesting that this dependence of betterness upon deontological concepts is at best only *one* possible interpretation of his logic argued against the superiority he feels his logic has over past evaluative logics of preference, given its supposed clarity. Interestingly, it is precisely the attitude of seeing his logic of preference in terms of "more fundamental" modal notions which separates Aqvist's attempt from that of von Wright's, in a striking way. The latter being more open to the acceptance of the viability of the notion of preference as standing alone, within a context of *ceteris paribus*.

In insisting upon a logic which is oriented upon a deontological basis, Aqvist sets forth the following requirements which aim at illustrating how the notion of 'better', symbolized as the binary relation 'B', is to be interpreted. In this connection he says that B holds between states of affairs p and q if and only if it can be maintained that any one of the following three conditions are fulfilled: "(i) p is obligatory and q is indifferent (neither obligatory nor forbidden), (ii) p is obligatory and q is forbidden, and (iii) p is indifferent and q is forbidden."[4]

Furthermore, Aqvist sees the preceding characterization of 'B' as conveying what he calls the "intuitive" import of the notion of better. This in itself is vital in appreciating the foundation from which he is launching his inquiry, namely one which relies upon an introspective recognition of "betterness," in a *derivative* sense of the term. Aqvist goes on to argue for the advantage of his

---

[3] Aqvist, Lennart., "A Binary Primitive in Deontic Logic," *Logique et Analyse*, 19, (1962), p. 90.
[4] Aqvist, Lennart., "Deontic Logic Based on a Logic of Better", p. 285.

approach in that he notes that if one were to evolve a logic of better on the basis of the good, bad and indifferent, i.e., on the basis of evaluative notions, then it is conceivable that two states of affairs may both be good, and yet one may be better than the other; in that it is "more good," so to speak. However, deontological concepts are not so readily given to ambiguity, in that the "obligatory" or the "forbidden" do not allow for degrees of attribution. From this perspective, Aqvist finds the approach he is employing sounder conceptually, as far as its providing a firm foundation for a logic of preference.[5]

Aqvist's enrichment of the propositional calculus begins with his interpretation of the expression: 'pBq', meaning that p is of higher deontic value than q, or alternatively that p is better than q. He introduces the following definitions as a means of extending the sense of 'B':

"pSq is short for -(pBq v qB--p)
Fp is short for -pBp
Pp is short for -(-pBp)
Ip is short for pS-p."[6]

The additional mechanism he uses from PC is the rule that if a is a tautology, then a is provable in his logic. Also, he retains the rule of detachment, which becomes the primary postulate of his calculus, (PC). The remaining postulates of his calculus are as follows:

"(Pl) pB-p & qS-q ⊃ pBq
(P2) pB-p & -qB--q ⊃ pBq
(P3) pS-p & -qB--q ⊃ pBq."

These postulates represent in symbolic form the three basic conditions announced at the outset of his analysis above, one of which must be met so that the binary relation B can hold between two states of affairs. Aqvist's postulate (P4) is essentially a disjunction of the previous three postulates:

pBq ⊃ (pB-p & qS-q) v (pB-p & -qB--q) v (pS-p & -qB--q)

Finally, the last postulate, (P5), is presented as:

(p ⊃ q)B-(p ⊃ q) ⊃ (pB-p ⊃ qB-q).[7]

Basically, (P5) assumes that if two mutually exclusive conditional states are such that one is said to be deontically better than the other, then it must be the

---

[5] ibid., p. 285.
[6] ibid., p. 286.
[7] ibid., p. 286.

case that their respective antecedents and consequents are such that their respective negations are deontically better than their opposite number. This appears to be indicative of the general view that if two states are such that one is better than the other, then any constituent part of the former state is such that it is better than any corresponding part of the latter state. The idea expressed above is also to be found in Aristotle's *Topics*, and thus in itself does not constitute an innovative insight. Furthermore, one can easily conceive of situations where this thesis simply would not hold. For though it may be obligatory that say p ⊃ q, from this it cannot be justifiably inferred that p in itself is obligatory, and that q is forbidden.

That Aqvist recognizes a relationship of basic dependence by his deontic logic of the better. i.e., the BD systems, upon modal logic generally, is again plainly evidenced by certain remarks he makes at the outset of his discussion in an earlier article, entitled 'A Binary Primitive in Deontic Logic.' First to be noted is the way by which he interprets all the fundamental definitions of the BD calculus in modal terms. Thus, running through the list of basic definitions he offers the following account:

"D1. Spq = def NABpqBqp
D2. Op = def BpNp
D3. Fp = def BNpp
D4. Ip = def SpNp
D5. Pp = def NBNpp"[8]

Clearly, one can observe in the above list how he conceives of the BD system as basically reducible to modal relations. Moreover, in describing the basic postulates for BD he proceeds to present in this earlier article two postulates which are virtually identical to postulates (P4) and (P5) in the later logic of the deontically better. Interestingly, in introducing P2. as the second postulate of his deontic system BD he writes: P2. CBCpqNCpqCBpNpBqNq, which is precisely what is meant by (P5) in the 1963 paper. Similarly, where he presents his third postulate he writes: P3. CEBpqAAKBpNpSqNqKBpNpBNqqKSpNpBNqq, which is semantically equivalent to (P4) in the latter essay.[9]

Aqvist proceeds to define a series of progressively more "powerful" deontic systems with "better," by replacing the fundamental rule of transformation in BDI with a more expressive rule. Thus, BD2 is obtained by replacing (Rl) in BDI, which says "if α ⊃ β is a tautology, then α B -α ⊃ β B -β. with the

---

[8] Aqvist, Lennart., "A Binary Primitive in Deontic Logic," p. 90.
[9] bid., p. 91.

stronger rule: (R2) α ⊃ β → α B -α ⊃ β B-β ; of course by defining α ⊃ β as "if α is obligatory, then β is obligatory," it follows that it is provable that α is obligatory, i.e., it implies that α is obligatory, i.e., α B -α. In essence, (R2) simply formalizes the meaning of the obligatory character of α and β, within the context of any implication relation which may be said to hold between α and β .[10]

System BD3 is derived in two steps. First, the following rule is added to BD2: "(R3) if α is fully modalized in the sense that all occurrences of propositional variables in α occur within some formula β B γ, then α B -α ⊃ α ." Secondly, there is the replacement of (P5) in BD2 by a new postulate (P5') (p ⊃ q)B-(p ⊃ q) ⊃ (pB-p⊃ qB-q)B-(pB-p ⊃ qB-q)

BD4 and BD5 are derived in succession as follows. Aqvist adds to BD2 the postulate (P6) pB-p ⊃ (pB-p)B-(pB-p). To get BD5 he adds to BD4 the postulate: (P7) -(pB-p) ⊃-(pB-p)B-(pB-p).[11]

While asserting the deontic character of the systems he is articulating, Aqvist proceeds to compare his approach to that of Soren Hallden's. The latter it is recalled emphasized more the evaluative dimension of the notion of better, and proceeded to present a logic which was thought to be immune to the vagaries which ordinary talk about values invariably involves.

Aqvist states in general terms that the Theory B developed by Hallden can be said to be contained within all the BD systems evolved above. However, (P4) in Aqvist's BD system is not provable in Hallden's Theory B. Aqvist notes that in the case of (P4) one encounters a situation where Hallden's infinite matrix M assigns a value to (P4) which is not designated according to M. Aqvist explains this lack of precise fit between Hallden's theory B and his own system as a matter which reflects the basic difference in the very notions of better, i.e., the evaluative account adopted by Hallden, and the deontic approach favored by him.[12]

Generally, any criticism of a modal logic of preference would be applicable first to a standard criticism of modal logic itself. This is to say that relative to the matter at hand, one's immediate reaction is to demand an explanation of just what the variables of Aqvist's logic are supposed to range over. This is a problem which arises from an assessment of modal logic itself, in that modal logicians can hardly ever clearly explain just what "propositions" are, over

---

[10] Aqvist, Lennart., "Deontic Logic Based Upon a Logic of Better", pp. 286-287.
[11] ibid., p. 287.
[12] ibid., p. 289.

which the variables of a modal logic are supposed to range. Furthermore, notions such as the "obligatory," the "forbidden," and the "permissible," are open to such a variety of interpretation, that it is often impossible to gain a precise sense of what these terms actually mean. In view of this, one questions the usefulness of such a logic, as far as its providing a simple and distinct basis for analyzing preferences. In summary, one seems to be hard pressed to accept the argument often voiced by defenders of modal logic that their logic has superior expressive powers, when compared to two-valued propositional logic.

The issue cited above is indeed a curious one especially because it is seen to arise where Aqvist attempts to compare his system with Hallden's. In the essay entitled, "A Binary Primitive in Deontic Logic," Aqvist notes that one point of difference between Hallden and himself is that the former is willing to accept in his Theory B wffs of PC, whereas this is inadmissible in his own BD system.[13] On the face of it, it would seem that Aqvist refuses to consider any comparability between propositional logic and his own deontic system. Yet he believes that a great portion of Hallden's system is expressible in terms of his own deontic logic of better. Thus, to deny on the one hand that the variables of his logic can range over propositions, contrary to the case in Hallden's Theory B, and still to maintain that Theory B can be expressed in terms of deontic theory BD, is to befuddle the sense of whatever Hallden may have meant to convey with his theory of better. Furthermore, Aqvist argues that instead of the rules of detachment, substitution, and interchangeability of equivalents used by Hallden, and which are also found in propositional logic, he will use an "analogous" rule of the interchangeability of equivalents, containing "suitable counterparts to the principles of inference" in Hallden's theories A and B.[14] Here again one may question the nature of the analogy between the rules in Theory B and his own BD system. How are these "counterparts," so-called, just as "suitable" as those in Hallden's Theory B? The question here is a pressing one mainly because if it is not adequately resolved one cannot go on to accept the affinity Aqvist sees between his system and Hallden's theories A and B.

Section II. *Chisholm and Sosa on the Logic of the "Intrinsically Better"*

For three important reasons Roderick Chisholm and Ernest Sosa provide a contribution of measurable historical value in the development of the area of

---

[13] Aqvist, Lennart., "A Binary Primitive in Deontic Logic", p. 94.
[14] ibid., p. 94.

prohairetic logic. First, they present a brief synapses of the work which had been done in this domain, going as far back as the early 1900's, — and beyond. In this they observe how in many cases in the past, general formal principles concerning preferences were only dimly perceived by analysts like Schwartz, Scheler and Brogan. Secondly, these authors provide important criticism of five preference-principles, popular at the time of their writing. Thirdly, they sketch a logic of preference of their own, dealing exclusively with the relations which hold between the "concepts" of the intrinsically good, bad, better and indifferent. This in itself stands in marked contrast to the work of von Wright, which focused upon preferences more broadly as the expression of a "special type" of proposition.[15]

Chisholm and Sosa base their criticism of other theories on cases from ethical hedonism, which say that the *intrinsically* good is that state of affairs which when realized manifests the greatest degree of pleasure, without violating any generally respected moral precept. Pleasure's presence is taken as being *in itself* the only thing which is good, and displeasure is that which is *in itself* bad. Both authors allude to Moore on this point, though they are less than specific on explaining the precise meaning of what clearly counts for the *hedonically* good. However, what is again striking at the very outset is that both begin with the premise that the logic of preference, or of better, must reflect a logic of a special type of discourse, namely *moral* discourse, dealing with the relationships holding between certain ethical terms, and not just with the logic of language in a broad sense.[16] This is certainly not the direction of inquiry found in Ramsey's efforts, nor is this precisely the case with Soren Hallden, nor with von Wright, nor with many other more recent figures. This admittedly new turn in emphasis is one which clearly brings forward the relevance of the study of formalizing preference, as a basic concept, centrally located within a philosophical inquiry concerning values. The only other approach which concerns itself with some modicum of the moral at this point in time is Aqvist's work as shown above; where emphasis is placed upon the obligatory, forbidden, and modally indifferent. However, the latter case is more closely connected with showing the possibility of interpreting a logic of preference in terms of a modal logic, rather than with considering the possibility of a logic of preference as basic to an understanding of moral discourse. Aqvist is thus more concerned with a kind of logical exercise. rather than with exploring the philosophical sig-

---

[15] Chisholm, Roderick M. and Sosa, Ernest., "On the Logic of the 'Intrinsically Better,'" *American Philosophical Quarterly*, Volume 3, 1966, p. 244.

[16] ibid., p. 244.

nificance of formalizing preferences within a context of discourse concerning values. The latter is surely of great importance to Chisholm and Sosa.

Perhaps the influence which induced Chisholm and Sosa in this particular direction originated from A.P. Brogan's little known essay in 1919, entitled "The Fundamental Value Universal." Though these authors come to reject some of Brogan's principles governing the better, it is nonetheless his central thesis concerning the importance of this notion which they come to adopt. Brogan sets his sights on showing how generally any discourse about value characteristics invariably involves the relation of "better."[17] Though he actually makes no specific allusion to the notion of preference as such, Chisholm and Sosa interpret his remarks on the centrality of better as reflective of what could also be said about preference, considered in terms of what is preferred intrinsically.

Specifically, Brogan's thesis is that in speaking about any value characteristic whatever, one must incur in some essential way some reference to a relation of better. He points out that whether one is taking into account: (1) the intrinsically "good;" meaning by this that the existence of A *as a fact* is good, or (2) what is best as opposed to what is worst, or (3) value simply as a universal; upon inspection all these concepts are definable in terms of that which is better. However, "better" itself is not definable in terms of any of these concepts. Herein lies the justification for Brogan's thesis that the better is *the fundamental value universal*.[18] Brogan extends his thesis to other value terms, some found in ethics, and some in aesthetics. In ethics, for example, terms such as right and wrong, ought and duty, etc. are all said to be reducible to a relation which expresses intrinsic betterness, however complex these expressions may be. In the case of certain key terms in aesthetics, like "beauty" for example, Brogan is found to refer to Moore, observing that since beauty is definable in terms of the intrinsically good, by consequence it is also reducible to a relation involving the intrinsically better.[19]

Apart from the drawbacks of Brogan's laconic mode of argumentation, his seminal realization of the centrality of better went unnoticed to Hallden and von Wright. However, in the hands of Chisholm and Sosa, Brogan's ideas became a conceptual foundation from which to launch their criticism of five prevalent preference-principles.

---

[17] Brogan, A.P.. "The Fundamental Value Universal", *Journal of Philosophy, Psychology and Scientific Method*, Vol. 16, (1919), p. 96.
[18] ibid., pp. 98-99.
[19] ibid., pp. 102-103.

In the first instance, they reject the principle which says that "p is good if and only if p is better than (or preferable to) non-p," i.e., (Gp ≡ pP-p). Their reasoning here is that one can have states of affairs which are better than their negations, but which are not in themselves intrinsically good. Thus, it is better according to the principle of propagating the greatest degree of hedonistic good that there be happy egrets, than that there be no happy egrets. However, there is nothing intrinsically good in there being happy egrets.[20]

In the same way, they reject the principle that p is better than q if and only if -q is preferable to -p, symbolically expressed as Bp ≡ -pPp. For the fact of there being no happy egrets is not in itself intrinsically bad. Again, it is worth reviewing the basic idea which animates Chisholm and Sosa's criticism of this and all forthcoming principles, namely that the intrinsically good state of affairs is precisely the one which alone makes the actual presence of pleasure a premium item.

The third principle is dismissed in a similar way. It states in symbolic terms:

pPq ≡ -qP-p

Substituting for the variables p and q one can see that though it is preferable that there be happy egrets (p) to their being stones (q), it does not follow therefore that there not being stones is better or worse than there being no happy egrets.

In a related manner, counter-examples are given for the principle:

pPq ≡ (p&-q) P (q&-p),

illustrating how this equivalence does not hold in every case.

The fifth principle: (p)(Gp v Bp v (-(pP-p) & -(-pPp))) is found to be false by reason of saying that there being no unhappy egrets (p) is neither intrinsically good nor intrinsically bad, though it is preferable that the state of affairs of their being no unhappy egrets is better than its negation.[21]

The authors make a point of noting how negative states, so-called, play no part in their conceptual analysis of preference. In their view, what is negatively good is no better than what is indifferent, and the latter is no better than what is negatively bad. This observation is surely consistent with the fundamental idea that one should consider the hedonically good as the realization of

---

[20] Chisholm, Roderick M. and Sosa, Ernest., "On the Logic of the 'Intrinsically Better,'" p. 245.

[21] ibid., p. 245.

the premier object of what is better, i.e., what gives pleasure, while violating no generally accepted moral precepts.[22]

One's understanding of the objects of preference as noted by Chisholm and Sosa above is properly put in perspective where their work is considered from the basis of the underlying conceptual foundations from which their analysis is launched. This is to say that presenting the philosophical tradition from which Chisholm operates, provides a means of seeing how one can have a logic of preference, whose objects are not necessarily physical entities existing or having to exist within a spatiotemporal plain. For where Chisholm's work is seen within the context of his evaluation of the contributions of Frans Brentano and Alexis Meinong, then the basic ideas of the "objects" of such a logic, as concepts, becomes more palatable. Moreover, the background of Brentano and Meinong provides the needed depth, which renders the intuitionistic approach employed by these authors more intelligible. Surely, Chisholm's detailed study of Meinong's work gives a greater degree of rationale as to why these authors adopt the intuitive approach toward evolving a logic of the intrinsically better. By contrast, Hallden and von Wright provide for scarcely a mention of the philosophical tradition from which they embark upon presenting their respective logics. Similarly, the conceptual basis of Aqvist's intuitive approach is also difficult to fathom.

Without doubt Brentano's views on the nature of preference are reflected in the kind of argumentation one finds in the analysis presented by Chisholm and Sosa. Quoting from the opening lines of Aristotle's *Metaphysics*, i.e., "all men naturally desire to know...," Brentano brilliantly alights upon the insight that all take the highest degree of pleasure in clarity of thought and judgment. Thus, it seems evident that one prefers that which provides such pleasure, and that pleasure is thus a necessary component of those things which are preferred. On the basis of the above, Brentano notes that the good is that which allows for the highest degree of pleasure, and the bad is that perverse state where pleasure is not preferred. Here the parallelism with the preliminary groundwork of the essay under scrutiny is unmistakable.[23]

To Brentano it makes no sense to speak of the intensity of pleasure in the context of preference, as if it were something subject to quantification. Rather, here one is dealing with the love which precipitates from the expansion of the

---

[22] ibid., p. 246.
[23] Brentano, Franz., *The Origin of Our Knowledge of Right and Wrong*, Routledge and Kegan Paul, New York, 1969, pp. 21-23.

intellect, through understanding. In summary, Brentano makes the remarkable statement that to prefer something is in essence an expression of one's love for the thing, because of the pleasure, i.e., end, it provides. Moreover, determination of the *correct* preference requires not only the knowledge of love and pleasure, but the experience of preference in past circumstances. These collectively provide for an awareness of what counts as proper preference. Preference is thus not divorced from the informed intellect or judgment. In this connection the concept of "better" is explained by Brentano as correctly preferring that which is good over that which is bad or indifferent. He notes that good allows for degrees of goodness, as does the bad, but truth and falsehood do not allow for degrees, since something is either entirely true or entirely false. Hence, the concept of better seen in terms of preference, presupposes a relative degree of good or bad, and is properly conceived in terms of what one would choose. This is not to deny, however, that one cannot prefer the true to the false, or that it would not be said that one prefers that which is good for its own sake to what is bad. The point Brentano appears to be concentrating upon is that the notion of better allows for the operation of relative judgments, and that this is only one of the possible contexts in which preference is found to be applicable. Having made the above observations, Brentano is careful to note that preference may also be involved with matters of practical choice, where no consideration of moral judgment is required. Furthermore, there may be preference in cases where *no choice* on the part of the agent is possible, as in preferring a sunny day to one with rain. This only underscores the broadness of the notion of preference as realized by Brentano, the core of which appears to underlie the analyses given by Chisholm and Sosa.[24]

Yet, there is more in Brentano which enables the reader to set Chisholm's work in sharper light. It should be recalled that Chisholm and Sosa stipulate at the outset of their analysis that their logic will deal with the formal relationships holding between intuited "concepts" of intrinsic good, badness, better, etc. Hence, since it is with concepts that their value calculus of preference will deal and not with propositions, or states of affairs. Concepts are taken as objects of judgment having no ontic reference other than being the idea presupposed by such judgment. Concepts in this specific sense are characterized by Brentano as "*synsemantic,*" and are suggested by Chisholm's opening remarks on how the proposed value calculus will be an effort to formalize the relationships holding between states which are good, bad, better or indifferent,

---

[24] ibid., pp. 26-27.

whether they are exemplified or not. Thus, the objects of the calculus are easily understood as resulting from an intuitive analysis of a particular insight of consciousness, i.e., one which concerns itself with judgment formation regarding proper choice.[25]

Though Brentano's work is helpful in seeing in broad terms the concept of preference as suggested by Chisholm and Sosa, it is Meinong's views on the concept of "*Objective*" which helps sharpen one's awareness of how the objects of their value calculus are to be taken. Without entering too deeply into an account of Meinongian *Gegenstandstheorie.* it should be sufficient to point out that for Meinong the *Objective* does not *exist*, but rather it *subsists*. This is basically the difference between *existieren* and *bestehen*. In judging that such and such is the case, what is said to be true is not of the object itself as an existent, but rather of the being of the object as a subsisting entity. In discussing Meinong's work on this point. Chisholm is found to say that other philosophers have used the expression "proposition" to refer to something of what Meinong means by *Objective*. However, the virtue of Meinong's terminology is held to be that it is relatively free both of linguistic and psychological connotations to which the term "proposition" is often exposed.[26] This point alone is quite significant in that it strongly suggests how Chisholm sees in Meinong's work the opportunity to attain a new conceptual foundation for his own analyses. For if it is the case that where he considers the objects of preference as Meinong thinks of *Objectives*, then he is trying to evolve a value calculus which is ontologically free of the issues which invariably emerge when one tries to assert something of that which cannot exist. The being of the *Objectives* is said by Meinong to belong to a higher-order of object, one which includes both those things which can or do exist, as well as those things which cannot or do not exist. Consequently, objects of preference would be of the type which could have non-being at the present time, and can be taken as subsisting or as having the same kind of non-being as the golden mountain, the round square, or the black swan.

Meinong cautions that one must not be unduly influenced by the actual in determining what may be said of the real. In this, there is a noticeable departure from Platonism, in that he is not saying that whatever is is in the world of

---

[25] Edwards, Paul., *The Encyclopedia of Philosophy*, Volume, Macmillan Company, & The Free Press, 1967, p. 367.

[26] Chisholm, Roderick M., *Realism and the Background Phenomenology*, The Free Press, New York, 1960, pp. 79-80.

actual physical reality, and *is* also, in an idealized realm, — though in a more perfect manifestation. For Meinong expands the notion of reality to include also those things which are impossible conceptually. Furthermore, it is important to point out that Meinong argues that there must be a distinction made between objects as presented by sensation, that is, things which are actual, and objects as presented by emotion, i.e., objects-being-worthy-of-interest, or objects-being-such-as-they-ought-to be. It is objects as presented by emotion or feeling which are "objects- having-properties." In other words, objects of desire have a certain quality belonging to them, in that they are worthy of our interest, or they are properly desired to be brought about. Thus, there is a judgmental component in objects as presented by emotion. The objects of preference then would be considered as subsistent, in that they are not actual, and since they are presented by feeling, they are objects-having-properties as well.[27] It is within this kind of notion of the preferable, one having strong similarities to Brentano's treatment of preference within a context of informed desire, which appears to underlie the conceptual basis from which Chisholm and Sosa are launching their analysis.

Apart from their criticism of the theories of others, which employ the notion of better as basic to the definition of good or bad, Chisholm and Sosa proceed to argue how in their view it is still the case that the better is the fundamental notion underlying good and bad, but for reasons other than those advocated in the previous theories. What the authors introduce as an innovation is the notion of indifference, defined as -(pP-p) & -(-pPp), related to the state of "neutrality," which in turn is defined as -Gp & -Bp, meaning that p is neutral where p is not good and p is not bad. As a result, they are able to introduce greater flexibility into their analysis by saying that a state of affairs does not fall only into the categories of the good, the bad, and the indifferent. In their framework there are seven possible states of affairs:

"(i) Gp and B-p
(ii) Gp and N-p
(iii) Np and B p
(iv) Np and N-p
(v) Np and G-p
(vi) Bp and N-p
(vii) Bp and G-p".[28]

---

[27] Brentano, Franz *The Origin of Our Knowledge of Right and Wrong*, p. 8 and pp. 11-12.
[28] Chisholm, Roderick M. and Sosa, Ernest., "On the Logic of 'Intrinsically Better'", p. 246.

Hedonists would reject the idea that a state of affairs could fit into both categories (i) and (vii). For example, in the former case one could not hold both that possible state p is good, and that its negation is also good. Moreover, in the latter case the hedonist would not hold that possible state p could be bad, and that also its negation is also bad.

The elements of the extended propositional calculus (EPC) with the two — place predicate constant "P," meaning "is intrinsically preferable to," are presented conventionally.

There is the usual specification for well-formed formulas with conjunction, disjunction and implication, etc., 'C' and 'D' are metalinguistic variables ranging over formulas of the EPC., and 'L' and 'M' are taken as metalinguistic variables ranging over formulas of the value calculus: VC.

At this point the authors identify a peculiar ambiguity in their logic, in that they conceived the proposed value calculus as dealing with variables which range over states of affairs. This, of course, is consistent with the basic premise presented at the outset which asserts that it is states of affairs, whether exemplified or not, which are properly said to be better or worse, and not propositions. Yet, in this very same value calculus, one wants to make assertions by way of stating axioms and proving theorems. However, one can do this only if use is made of the logical machinery developed in the extended propositional calculus. Consequently, herein lies the systematic ambiguity in Chisholm's and Sosa's analysis. They illustrate the problem by referring to the following logical expression in their value calculus: ((p v-q)Pr) v -(sPt). In their first occurrence '-' and 'v' are "exemplification functional operators," working on *states of affairs* yielding compound states of affairs, e.g., ((pv-q)Pr). However, in the second occurrence,'-' and 'v' operate as truth functional operators, ranging over *propositions*, relating (p v -q)Pr to -(sPt) by way of disjunction.[29]

The authors recognize the problem and admit in essence to it being a difficulty which must be tolerated in their presentation of VC. Clearly, they identify the problem in a more direct manner than von Wright, who argues that somehow the objects of preference are "proposition-like" entities, which in itself is an attempt to steer a middle course between a commitment to objective physical states of affairs, and linguistic entities, i.e., propositions. The sense of "proposition-like" entity, of course, leaves a great deal to be desired, and it is not at all to the detriment of the attempt by Chisholm and Sosa that they see the problem as present, and as formidable.

---

[29] ibid., p. 247.

In a sense, it could be said that the above difficulty is a manifestation of the same type of issue encountered by Ramsey and Davidson, namely that of questioning the nature of the relation between the theoretical model and the thing or object analyzed. Can there be a precise fit between model and object of inquiry, or is one forever estopped from bridging this gap? This question is now found to emerge repeatedly in the assessment of works by Hallden, von Wright, Rescher, Martin, and Jeffrey. It is perhaps the central problem which lurks in the background of the entire history of this area of philosophical inquiry.

The essential definitions of the system (VC) are as follows:

Sameness in intrinsic value is defined as:

D1. pSq FOR -(pPq) & -(qPp)

Intrinsic *indifference* is defined as:

D2. Ip FOR -(pP-p) & (-pPp)

Intrinsic *neutrality* is defined as:

D3. Np FOR (∃ q)(Iq & pSq)

Intrinsic *goodness* is defined as:

D4. Gp FOR (∃ q)(Iq & pPq)

Intrinsic *badness* is defined as:

D5. Bp FOR (∃ q)(Iq & qPp).[30]

The axioms presented are not wholly without controversy. Those of the less questionable variety are the simpler ones:

(A1) (p)(q)(pPq ⊃ -(qPp))
(Non-contradictory relationship)

(A2) (p)(q)(r)[-(pPq) & -(qPr)] ⊃ -(pPr)]
(The transitivity of the preference-relationship)

The third axiom presents some difficulty since it assumes that two states of affairs are the same intrinsically if individually they are not preferable to their respective negations., i.e., they are intrinsically indifferent.

(A3) (p)(q)(-(pP-p) & -(-pPp) & -(qP-q)
 & -(-qPq)) ⊃ (-(pPq) & -(qPp))[31]

It is specifically the notion of "intrinsic sameness," suggested in (A3) above, which, is suspect at this point. Surely, this cannot be taken as an identity relation, but as some kind of an intuitively recognizable similarity of one sort or another. Considering this matter strictly in terms of the conceptual framework

---
[30] ibid., p. 247.
[31] ibid., p. 247.

adopted by the authors, it is to be noted that their presentation of the concepts of intrinsic goodness and badness is taken as meaning that such states are in themselves good or bad relative to whether they make possible the precipitation of pleasure. However, it could be argued that since the determination of pleasure is a highly subjective judgment, the pronounced emphasis upon the very intuitive recognition of whether some states promote or impede pleasure works against the idea that an ostensive sense of sameness is being or can be conveyed. In reality the "intrinsic sameness" can only be taken as the degree of pleasure which is *judged* to have accrued from some states of affairs, through the assessment of an independent observer. The entire framework of analysis proposed, therefore, is subjectivistic to such an pervasive degree that it is difficult to see how it can constitute the objective foundation needed for a logic. This, however, may only be symptomatic of the phenomenological background from which their analyses emerge.

Their fourth and fifth axioms can be rendered into one, as is observed by these authors in another article, "Intrinsic Preferability and the Problem of Superogation." Here Chisholm and Sosa state their last axiom as follows,

(A4) $(p)[[(q)(Iq \supset pPq) \vee (q)(Iq \supset qP\text{-}p] \supset pP\text{-}p].$[32]

The rules of inference to be employed are:

(R1) Relative to EPC and VC, where a wff of VC is substituted for each of the propositional variables in a theorem of EPC, the universal closure for what results, with respect to each of its variables, is a theorem of VC.

(R2) In this rule Chisholm and Sosa assert that the reverse substitution mentioned in (R1) holds as well.

(R3) Modus Ponens

(R4) The fourth rule is somewhat complex. Essentially, it can be stated as follows:

(a) C is a theorem of VC,
(b) D ≡ E is a theorem of EPC,
(c) F is a well-formed formula of VC,
(d) if F and C differ in that F has E wherever C has D, and F has at the beginning a universal quantifier for any variable in E which is not in C, *then*
(e) F is a theorem of VC.[33]

---

[32] Chisholm, Roderick M., and Sosa, Ernest., "Intrinsic Preferability and the Problem of Superogation", *Synthese*, 16, (1966), p. 324.

[33] Chisholm, Roderick M., and Sosa, Ernest., "On the Logic of 'Intrinsically Better'", p. 247.

The authors present a total of forty-three theorems, ranging from very simple relations, e.g., (T6) (p) -(pPp), to complex theorems requiring eleven or more steps, e.g., (T21) (p)(pP-p ⊃ (Gp v B-p)). To present all these theorems would not seem to be adding to the elucidation of the Value Calculus. However, some representative examples may be helpful, at least by way of providing a suggestion of the applicability of this logic. Thus:

(T3) (p) [(q)[(Iq ⊃ pPq) v (q)((Iq ⊃ qP-p)]⊃ pP-p)]
        (A4), (D2);
(T15) (p)((Gp v B-p) ⊃ pP-p)
        (T3), (T12), (T14), (D4), (D5);
(T18) (p)((Np & Nq) ≡ Ip)
        (T16), (T9), (T17); etc., etc.[34]

Section III. *Bengt Hansson on Fundamental Axioms for Preference-Relations*

Within the now emerging tradition of thinkers whose approach to the development of a logic of preference involves intuitionistic analyses which avoid any unnecessary restrictions by specialized contexts or subject matter, appears the contribution of Bengt Hansson. As already noted his effort comes after von Wright's monumental contribution, and thus it cannot be discussed apart from referring to some of what the former had clarified on a conceptual basis. However, it is nonetheless a view of preference which also incorporates a relevant discussion of Hallden's *Logic of Better*. Thus, Hansson serves as a convenient bridge between Hallden and von Wright. In a sense it could be said that his work attempts to incorporate Hallden's investigation, and also serves as a way of introducing von Wright's work through indirect criticism.

Unlike most of his predecessors, Hansson is interested in presenting a logic of preference which reflects what is meant in "everyday talk" when the expressions "is preferred to" and "is better than" are used. Thus, there is a pronounced shift in his work in contrast to preceding efforts, at least with respect to the recognition that a formalization of preference-relations must receive its cogency or general signification from the discourse one encounters in ordinary usage.[35] This is a far cry from the attitude one finds toward ordinary

---

[34] ibid., pp. 247-248.
[35] Hansson, Bengt., "Fundamental Axioms for Preference Relations," *Synthese*, 18, (1968), p. 423.

talk in Davidson's early work in this area, as it is also a change from Hallden's reference and use of the ordinary language context. Hansson also expands the basis for the relevance of a logic of preference somewhat further than Chisholm and Sosa. The latter sought to formalize preferences solely within the specialized areas of moral discourse, whereas Hansson sees the logic of preference as having applicability in the vastly more complex sphere of ordinary discourse, which would thus require absolutely no contextual restrictions. The only other investigator who comes close to Hansson's recognition of the need to evolve a logic of preference as it is presumed to operate in everyday discourse is Richard Jeffrey. The latter, however, eschews any use of modal logic, and relies exclusively upon a quantificational approach to the problem, much in the tradition of Ramsey, Rescher, Martin, and Davidson. Moreover, one sees here as well a refusal to be restricted by any "intrinsic" interpretation of preference, as encountered in von Wright. For Hansson takes his formalization of preference axioms as applicable on a cross-contextual basis. This is to say that he sees a need for a single and distinct logic of preference, having application in business management, hygiene, economic theory, jurisprudence, etc. Furthermore, if, according to Hansson, one is to achieve this formalization, ultimately having use in a wide variety of areas, one must first establish what he terms "the preference field." This is a range of preference-relations having the most pervasive applicability. One specifies this field by determining in each *particular area* only those preference-relations which are among the elements of well-defined sets.[36]

The ontology Hansson envisions for his calculus is one where 'P', expressing the binary preference-relation, holds between "things, classes of things, situations, decisions, or performances."[37] In this way he hopes to achieve a broader scope of utility than past efforts which had the binary connective operating only over *propositions*, expressing the above type entities. Clearly, this can be taken as a direction of inquiry having some variance from the approach found in Chisholm and Sosa, who see the reference to propositions as an unavoidable necessity, if one is going to make use of the functions of the propositional calculus. This is no doubt also a departure from von Wright's manner of alluding to the objects of his logic as "proposition-like" entities; necessarily reflecting linguistic units of some sort, though not clearly identified as such.

---

[36] ibid., pp. 424-425.
[37] ibid., p. 423.

At the very outset, Hansson endeavors to explain the nature of the tenuous relation holding between his proposed axiomatization of preference-principles and propositional logic generally considered. As seen above, he makes absolutely no restrictive assumptions concerning the interpretation of the objects of his calculus. The determination of what 'p', 'q', 'r', etc., refer to will be specified only whenever a concrete example is given. More interestingly, the composition rules and operators of the propositional calculus may not be defined in the preference field, but if they are so defined they are not taken as having the *same* meaning as "similar-looking tokens" found in his axiomatized logic. However, Hansson goes on to stipulate that though the formal properties of these operators and rules are taken as being the same as those in the propositional calculus, in some very vital though unexplained sense they are also said to be independent from the latter. Hansson's aim here is to secure a logic which is as far as possible uncommitted to any context which would limit its scope, even when that context is that of propositional logic. This is no doubt consistent with his conception of the logic of preference, and to this extent one can appreciate his insistence that the objects of this logic are not to be limited.[38]

Hansson also proceeds to see his logic of preference as having applicability not only on a cross-contextual basis, but it is even useful beyond the constraints of the intrinsic/extrinsic distinction first identified by von Wright. This is to say that von Wright's observation that one can deal with preference extrinsically, i.e., in terms of the limitations imposed by certain external environmental conditions, does not impose a constraint on Hansson's logic. In Hansson's view, intrinsic or extrinsic considerations are simply different manifestations of context working upon the articulation of a logic of preference, and as such this should not play a determining role in what that logic should ultimately be in its formal character. This supposed independence from the influence of context is believed to allow for applicability in a wide variety of areas: aesthetics, ethics, science, value theory, etc., etc. The extent to which Hansson is willing to go in claiming that his will be a logic totally removed from any dependence upon context is further evidenced by his insistence that his logic is applicable in cases where preferences are considered either normatively or descriptively. In this he is attempting to argue that his logic should be able to straddle the two basic approaches emerging in philosophical attempts at evolving *prohairetic* logics. The one, pioneered by Hallden and developed by von Wright and

---

[38] ibid., p. 425.

others, being the normative approach to a logic of preference. The other, the descriptive approach, is initiated by Ramsey, and furthered by Davidson, Bohnert, Rescher, Jeffrey, and many others.[39]

Apart from his insistences that his logic is supposedly free of any limitation, a problem definitely appears where he says that the operators and rules of composition of his logic may at best only be "similar-looking" to those in the PC. Somehow, Hansson wants to draw a subtle distinction between the formal properties of these rules, and their respective meanings within his own logic of preferences. Yet one cannot but questions the tenability of this distinction since the sense of 'v', '-', '.', etc. is derived from understanding the specific functions of these symbols in the defined field of propositional logic. Consequently, their meaning *is* their functional use in the PC, and there does not seem to be a sensible way of distinguishing between the formal sense of these symbols, and any other special type of meaning they may derive from within Hansson's logic. Hansson's position here seems to be extraordinarily obscure. For it is surely evident that it is not ordinary discourse which enlightens the sense of '.' and 'v', as employed by Hansson. Rather, it is our knowledge of propositional logic which makes these terms intelligible to us. What Hansson means by saying that there are "similar-looking" aspects between compositional terms in his logic and their apparent counterparts in propositional logic is simply not explained. Nevertheless, this point is quite crucial to Hansson's entire thesis, since it serves to severely undermine his view that he can derive a *rigorous* logic of preference which is also *so* context-free, it is even unrestrained by the conceptual limitations apparently inherent in propositional calculus.

Furthermore, one of Hansson's operating assumptions is that he will present a logic of preference which will closely parallel the sense of the relation of preference in everyday discourse. With his insistence, however, that the operators and rules of composition which this logic deals with are only "similar-looking" to the ones in the propositional calculus, one is faced with the dubious thesis that somehow ordinary discourse has embedded within it rules of logic, akin to those encountered in propositional logic. Here, in keeping with Hansson's opening remarks, one must be careful to say that these are not the rules of propositional logic, but rather they are logical rules which are "similar-looking" to those of propositional logic. Assuming that it makes sense to speak of such "rules," this is surely quite beyond what the surface meaning of spoken discourse seems likely to allow, if the Chomskean thesis is for a moment

---

[39] ibid., p. 424.

disallowed. It is evident that Hansson's position at this point is far removed from the common belief that formal logic has the primary and invaluable function of clarifying the ambiguity that often infects ordinary discourse.

In presenting his axioms, Hansson stipulates that 'P' will not be taken as a primitive expression of the preference-relation, but that 'R' will serve this purpose. He sets down 'R's meaning as either "'at least as good as'", "'equally good or better than,'" or "'is preferred to or considered equal to'." Without proceeding further, there seems to be an evident ambiguity already built into the meaning of 'R'. Surely, to use one symbol for both the notion of x being as good as y, and for saying that x is preferred to y, is to put too great a degree of flexibility in the significance of any one specific symbol. Evidently, it could be argued that one car may be *as good as* another, for the sole purpose of excluding the possibility that one automobile is to be preferred to the other. Hence, it is possible that the meanings Hansson sets forth for 'R' can be incompatible, if not contradictory, *within the very context of ordinary* talk.[40] Yet out of fairness to Hansson, it must be recognized that what he endeavors to do is to use 'R' in the broadest sense possible, so that as his analysis progresses he will be able to narrow that sense of 'R' through logical relationships. In a word, 'R' manifests the greatest possible degree of generality, from which the notion of preference will emerge as a relational term of a degree of specificity greater than 'R'. However, though Hansson's motives are intelligible, the fact that he allows 'R' also to mean "is preferred to" at the very outset of his inquiry, hampers his argument that the preference-relation 'P' will somehow be *derived* from the 'R'-relation. For the latter already is taken as conveying "preference" in some very broad and unspecified sense. This is a central difficulty which will trouble his account of the relations of preference and of sameness.

In articulating his logic of preference, the first axiom asserts the transitivity between p and r. Hence,

A1 pRq & qRr → pRr.

As it stands, axiom A1. is meant to bring out the varied senses of 'R' in the relation of transitivity. This can involve the notions of "better than," and "is preferred to." Thus far there is no problem. However, the above axiom could also be read differently, so that 'R' conveys its alternative and quite different sense of being "as good as" or "considered equal to." In the latter case, 'R', meaning: "is the same as" in the relation of transitivity, can be interpreted as involving the more specific and, therefore, more precise relation of identity.

---

[40] ibid., p. 426.

The point here is that while the transitivity relation works where one considers 'R's meaning on an individual basis as in: "is as good as," "is better than," "is the same as," "is preferred to," etc., etc.; when one reflects upon the sense of 'R' generally, one must conclude that a very different thing is talked about when the meaning of 'R' in the transitive relational mode deals with p being the same as q, than where one says that p is preferred to q. In the latter case, transitivity involving sameness is surely less problematic than transitivity involving preference or goodness, or betterness. Thus, given the total scope of meanings conjointly which 'R' is said to have, Hansson's first axioms can be ambiguous, for this term needs a considerable amount of unpacking.[41]

His second axiom is taken as expressing the totality of the R-relation:

A2.   pRq v qRp

Axiom A2. is crucial for Hansson, and he identifies it as the "Axiom of Comparability."[42] Yet, as if sensing the possibility for ambiguity in his second axiom, Hansson observes that the plausibility of A2. depends upon the tightness or the homogeneous character of the preference field. However, in doing this he is making the clarity of his logic reside more upon the context of what is preferred, rather than upon the rigor of the logic itself. This gives to his analysis more of an extensional significance than he perhaps intends, given that he wants his logic to be as context-free as possible.

The first theorem derived by Hansson is gotten from A2. by substituting p for q. This is Theorem 1: pRp. Moreover. when A1. is combined with Theorem 1, the meaning of 'P' emerges as a "preorder" relation. Presumably, this means that P is in some sense to prioritize p in relation to q, which is not what 'R' is designed to do by itself. Hansson goes on to say that when the conditions for A1.,A2., and Theorem 1. are fulfilled, one has what he terms the *total preorder relation*, which he calls the "preference-relation." It should be observed here that this preference-relation is derived in a way which proceeds from successive orders of generality, so that one begins with the very general relation of transitivity in A1., and then on to the more restrictive Axiom of Comparability, A2., to finally theorem 1., which introduces pre-ordering.

An additional definition is given which sets forth the *partial preference-relation*, as a relation satisfying only A1. and the first theorem. Apparently, Hansson feels that a case can be made for the distinction between *a partial and a*

---

[41] ibid., p. 426.
[42] ibid., p. 426.

total preference-relation. Nonetheless it is interesting to inquire into the basis of the distinction between these two kinds of preference-relations. When is a preference-relation "partial"? To Hansson the answer resides in whether or not the Axiom of Comparability has been introduced. In its absence, the preference-relation is only partial. However, on his own admission the sense of the second axiom depends upon the uniformity of the context one is dealing with. Thus, it seems that at best the difference between total and partial preference can only be a distinction based upon context, if at all. Thus, also it is not clear that the difference between total and partial preference-ranking is a logical one, as opposed to a purely extensional one, based upon one's assessment of the unified quality of the preference field in specific situations. Ironically again, what seems to be happening with certain key concepts in Hansson's logic is that the very contexts from which he is attempting to liberate his logic become the sole source of understanding what his basic logical concepts are supposed to mean.

Hansson's third definition asserts that "pPq" is just another way of saying: pRq & -(qRp). Actually, the definiens in this case is supposed to be removing from the significance of 'P' any connotation of the: "at least as good as," "considered equal to," and "equally good." For Hansson, expressions such as the above illustrate that the significance of 'P' is definable in some derivative sense from R, with the use of certain logical operators.

Hansson's rendition of the relation of "same as" is again meant to illustrate how an essential relational term of his logic can be derived from the formal implications of 'R'. Operating on the basis of the varied meaning of this term, he proceeds to present his fifth definition as: "pSq is short for pRq&qRp". He maintains that 'S' is not only to be interpreted negatively as the denial of the existence of a preferential-relation between p and q, and q and p, but that it should also be given a more positive rendering, in the manner suggested by: "that a relation of sameness is said to hold between p and q. 'S' is taken as expressing a "positive degree of equality," and it is tantamount to the meaning of 'I' (indifference) in other logics."[43]

One could question whether there is anything in the logic itself which would warrant a "positive" as opposed to a "negative" interpretation in the above case. It would appear that Hansson's concern here is motivated by a sensitivity to the fact 'R' does have a very broad meaning, and that perhaps the preferential sense which is allowed for it at the outset of his discussion would

---

[43] ibid., pp. 426-427.

get in the way of the clarity which he hopes to secure for his S-term. Moreover, he would also want to avoid the troublesome position of having to explain contrary preferential states as expressions in his logic. Yet, his insistence that the positive interpretation be employed additionally does not necessarily eliminate the issue here. If anything, his stipulation on the "proper" interpretation of 'S' makes his logical analysis that much more obscure, and again points to the problem inherent in the very broad sense given to 'R'. Finally, in this regard one may ponder the nature of the "degree" of equality 'S' is supposed to express. Is this equality strong enough to be construed as an identity? Can 'S' be said to symbolize an identity-relation between states, as well as between things, and processes? If such is not the case, then it is no less difficult to understand how 'S' applies in the sense of sameness to things, as well as to situations and decisions.

Here it is useful to recall all that 'R' is intended to relate. This is to say that Hansson considers his logic as applying to: "things, situations, decisions, performances, etc., etc." The sheer diversity of ontology which is being alluded to here, namely states, processes, and actions, would seem to demand some specification of an event logic of some sort prior to his analyzing preference sentences. Yet Hansson simply proceeds on the assumption that 'p', 'q', 'r', ... etc. can be referring to any of the above kind of events without having any fundamental effect upon the clarity of his analysis. Surely, hindsight has provided us with an appreciation of the difficulties when attempting to hold the view that one prefers an object in the same way in which one prefers a statement being true of an object. There are differences of meaning here which demand careful attention, if clarity is to be secured on a semantic level.

The problems with Hansson's logic seem to persist, since he goes on to claim that his second theorem presents a simpler way of expressing the P-relation as: pPq ↔ -(qRp). This is, of course, a matter of debate, since the biconditionality claimed here by Hansson is not supported by the sense of what was first set forth as the meaning of 'R'. If such a biconditionality is to be upheld, then from the original sense of 'R' it could be maintained logically that p is preferred to q, because by negating that q has any relation of sameness, preference, betterness, or goodness, with respect to q, it *neccessarily* follows that p is preferred to q. On the basis of what is this biconditionality supported? Hansson argues that this second theorem is derived from the definition of 'P' as pRq&-(qRp), and the application of the Comparability Axiom. True to his method of operation, one sees here how Hansson is attempting to secure a formal concept in his logic by first stating a very general relational idea, e.g., pRq&-(qRp), and then restricting the range of that idea by a related but more specific notion, e.g., pRq v qRp, the Axiom of Comparability. However, at best the biconditionality here is applicable only in cases involving total preordering, since he uses Axiom 2. which does not cover partial ordering. Thus, the second theorem is restricted, and since it relies upon the Comparability Axiom, it requires a reference to context for its ultimate signification. The picture appearing here again is of a logic which is not unencumbered by the weight of its primitive terms.

Difficulties emerge as well where Hansson attempts to relate his axiomatic system to other attempts, especially to that of Soren Hallden's.[44] Hansson argues that Hallden's two primitive signs, B and S, are such that B corresponds to the P-term used by Hansson. The nature of the correspondence here is difficult to accept. First, recalling the exegesis of Hallden's work previously, it should be pointed out that unlike Hansson, Hallden does not evolve a logic of preference. The formalization of the better is not a formalization of preference expression, since the praxiological component necessary for the formalization of preferences is simply not taken into account by Hallden, much to the detriment of the success of his logic as understood from von Wright's criticism of his work. Consequently, it is not clear in what way Hansson's symbol P is *like* or *"corresponds to"* Hallden's B, the symbol for the relation of better. Secondly, and of equal importance, is the point that whereas Hansson attempts to develop a logic which is flexible enough to capture the open-texture of "ordinary talk" about preferences, Hallden was seen to deemphasize the importance of ordinary discourse as a source of gaining insights into the formal relations regarding the better. In fact, Hallden spurns any reliance upon ordinary discourse, choosing instead to evolve a logic of better which is comparatively "purer", and thus not susceptible to the ambiguities found in ordinary discourse. In view of these two basic differences, as well as the fact that Hallden's ontological presuppositions are less inclusive than Hansson's, it seems that the issue of the nature of the proposed correspondence between Hallden's symbol B and Hansson's P symbol is indeed difficult to resolve.

The above criticism serves to pose the further question as to the tenability of Hansson's claim that his axiomatization of preferences is "powerful" enough to assume most of the axioms in Hallden's theory B. Basically, it seems

---

[44] ibid., p. 427.

Hansson's claim suffers because of the opacity of key concepts which underlie his axioms. This was seen to be especially true of his 'R' term, which simply put is made to mean *too* much. Moreover, similar objections can be raised to the theorems Hansson evolves, i.e., the fundamental terms by which these theorems are derived are open to such a wide variance of interpretation, that his logic lacks the rigor necessary to serve a sufficiently productive purpose.

The remainder of Hansson's discussion is devoted to a critique of past efforts at deriving logics of preference, and the relative merits of these attempts in view of his own contribution. Hansson grants that his survey results in the negative conclusion that up to the time of his writing no attempt, including his own, has been able to carry the logic of preference beyond the "trivial part" reflected by axioms A1. and A2. above. Hansson's concluding remarks are an indictment directed not so much toward the abilities of his predecessors, but rather it is aimed at what he takes to be the expressive poverty of propositional logic. The latter is seen as being unable to capture the nuance of meaning which discourse about preferences seems to entail. His concluding positive comment is set on saying that modal logic of some as yet undiscovered but more powerful sort should provide the means of capturing, through the application of modal operators and probabilistic concepts, the proper formalization of preferences.[45] Ordinarily one would think that Hansson would devote a great deal of attention to the kind of approach taken by Aqvist in "Deontic Logic Based on a Logic of Better." Surely, his concluding remarks on the promise which a sufficiently powerful modal logic would have in this area precludes his assenting to Aqvist's work. Hansson, however, scarcely mentions Aqvist even once, and seems totally oblivious to what the latter had contributed.

Accepting aspects of Hansson's assessment of previous efforts in this area entails the adoption of his own axiomatization as more unambiguous than he himself believes it to be. However, in spite of the difficulties, what he sets out to do is truly commendable, given the historical tradition from which he is attempting to depart. Unlike any other investigator before him, he is concerned with a logic of preference in the broadest sense possible. One which will somehow accommodate the plasticity of the idea of preference as it is encountered in everyday discourse. This emphasis on "ordinary talk" as a criterion for the development of a logic of preference is a new direction in this area of inquiry.

---

[45] ibid., p. 441.

Yet, his insistence upon the total independence of such a logic from any possible context creates — in a manner of speaking — a diaphanous entity, one which virtually defies interpretation. What seems to be lacking here is the kind of practical appraisal of the situation found in von Wright. There seems to be a need for realizing that the implementation of propositional logic is an invaluable tool by which to evolve a prohairetic logic, and that logical analysis and practical concerns should not be perceived as two mutually distinct and totally unrelated domains.

Chapter 7.

Von Wright's Logic of Propositions Expressing Preferences

Georg Henrik von Wright's work on the logic of preference represented in its time the most thorough treatment of this area of twentieth century philosophy. Without question, von Wright performed the invaluable task of carefully defining the range of a logic of preference, considered in an *intrinsic* sense, demarcating the extent and limits of its scope. His understanding of the complexity of the theoretical issues with which this logic deals introduced a wholly philosophical attitude to the enterprise of developing such a logic, in marked contrast to prior efforts which relied heavily upon techniques developed in economics and related areas. Von Wright's early effort was to affect almost every subsequent attempt in this particular area of philosophical logic.

Von Wright does not argue that since values are inherently "alogical" they must of necessity deal with the practical, whereas by contrast logic, being essentially abstract, must deal only with the formal. This type of argumentation he says is founded upon the mistaken identification of logic *solely* with the theoretical, and value studies *only* with the practical, and that, therefore, these two areas are incompatible. Such reasoning also fails to consider the usefulness of propositional logic as an analytical tool which clarifies and explains the essential logical form of value claims in ordinary discourse. Thus, the logic von Wright endeavors to present is "a formal system of a basic and (logically) rather 'primitive' type of valuation."[1]

At the outset, it is best to be clear as to why von Wright enters into the enterprise which will culminate in his attempted logic of preference. Basically, in studying the formal properties of preference he is interested in evolving a logic of value, in the broadest terms. This is to say that his ultimate concern will be to show how from a logic of preference one could proceed to a logic of the different degrees of good and bad, and possibly also to a logic of good and bad in an absolute sense. Thus, von Wright is driven to the formalization of a gen-

---

[1] Von Wright, Georg Henrik, *The Logic of Preference*, Edingburgh University Press, 1963, p. 9.

eral theory of value. Moreover, he distinguishes what he will attempt to do from the logical study of norms as this has evolved into Deontic Logic. It is interesting to observe that, though von Wright was a major figure in the development of deontic analysis, he recognizes certain "grave insufficiencies and ... errors" in this approach and elects to pursue the analysis of the concept of preference which in a pioneering way will stand as perhaps an axiological counterpart to Deontic Logic.[2] Also, it is important to keep in view that von Wright's approach is entirely intuitionistic. This is to say that he proceeds on the assumption that one can intuit the logical form of the concept of preference as somehow a goal which when accomplished will be acceptable to all. This may or may not have been the influence of Soren Hallden on von Wright since, as has been seen in pursuing the logic of the better and same, Hallden is similarly interested in arriving at his logic in a puristic form, by the use of intuition. However, though it will be shown that von Wright recognizes the relevance of Hallden's work to his own, the former's departure from a purely deontological study of preferences indicates a separation at least in approach between Hallden and himself.

Indeed, as this study of von Wright's contribution progresses, it will be of consequence to note how little of Hallden's work he will be able to accept, (if anything at all) as relating to a logic of preference. This is the insight which occurs to von Wright in his 1972 essay, and his concerns about the conceptual implications of his later attempt will be brought forward as the account of his work in 1963 unfolds. Yet even in his revisitation of the problem in the later essay, von Wright preserves the belief that the logic of preference should be pursued on the basis of an intuitive analysis of the concept, and that the most essential sense of preference is preference considered *intrinsically*.

Apart from his incisive differentiation of deontology from praxiology and anthropology, alluded to previously in other contexts, von Wright - along with H.G. Bohnert and R.M.Martin - was also one of the first to recognize that a logic of preference should take into account persons, segments of time, and subjects.[3] Consequently, he sees preferences as necessarily dealing with an inter-crossing of axiological and anthropological factors. For this reason as well, early in his pioneering book: *The Logic of Preference*, he concentrates upon *preferential* states of affairs, rather than upon simply the formal semantics of discourse involving preferences. However, since he is primarily interested in

---

[2] ibid., pp. 8-9.
[3] ibid., pp. 12-13.

securing an intuitive logic of preference, he does not dwell upon explicating the extensional features of such states of affairs, as is found in the work of R.M.Martin for example. This is understandable, in part, since his overriding concern is to secure a formalization of preference-relations as a means of elucidating a general theory of value. Thus, von Wright's starting point, so to speak, is not from the ground up, i.e., he does not proceed from an observation of empirical phenomena from which he will build a logic of preference. The latter is the kind of approach Martin will favor with his extensional and pragmatic analysis. Rather, in reading von Wright, there is assumed already in the background of his analysis a conception of good and bad and of better and worse, which is intuitively self-evident. In other words, von Wright begins with a presupposed theory of value, i.e., the intrinsically liked, as intuitively evident, and evolves a logic of preference as a means of explicating the logical structure which this theory should have.

Precision for von Wright is not to be found in the specification of the extension of preferential states of affairs, but in determining the logical structure of relations involving preference expressions. Interestingly, no two attempts in this area are so diametrically opposed to each other methodologically as those of von Wright's and Martin's. The former is concerned generally with possible-world states in dealing with preferences and suggests that there are several logics of preference which can be attained through proper investigative procedures. Martin concludes that preference is tied to a logic of specific choice, and that there cannot be a single all-inclusive logic of preference, but rather there may be different logical analyses, depending on the state of things one is examining.[4] In the course of this inquiry, one will observe the sharp contrasts between the works of these two thinkers, and the implication this has for this study philosophically.

To understand more thoroughly the sense of "state of affairs" which von Wright employs in the development of his logic, one can refer to his later article "The Logic of Preference Reconsidered," where he elaborates upon the difference between states and things. In this essay he sees states of affairs as "proposition-like" entities which can be negated and handled in accordance with the laws of propositional logic.[5]

Thus it seems natural to say that where x is a better state than y, one can also mean to say that y is worse than x. Put in symbolic terms one has xBy = -

---

[4] Von Wright, Georg Henrik, "Logic of Preference Reconsidered," *Theory and Decision*, Vol. 3, (1972), p. 141.

[5] ibid., p. 143.

(yBx), which indicates for von Wright that the concept of the better is interdefinable with the concept of the worse, and that the two, by means of the operation of negation, are essentially one concept.[6] Thus, it is "the better" states of affairs *as* "expressions" of states of the world, *in contrast to* "thing-like" objects, which will become the object of von Wright's calculus of preferences. In the same essay he goes on to argue that preferences between states of affairs seem *more basic* than preferences between things. His argumentation here appears to be somewhat laconic, yet it suggests in a somewhat general way that upon analysis preferences concerning things involve *implicit* references to persons and occasions within segments of time.[7] Thus, preferences involving states of affairs become evident beneath the surface of analyzing preferences among thing-like objects, so to speak. For this reason, to concentrate upon the former would, in essence, also cover any logic dealing with the preference of particular things.

Preference is also related to betterness and choice. As noted above, preference is involved with both axiological and anthropological notions. Thus, preference's relation to betterness, taken in an *intrinsic* sense, is where it could be said that one has a personal liking (preference) for the better thing, object, etc. Though there is also an *extrinsic* sense of betterness, involving the utility or range of actual effects of specific kinds of action, it is the sense of preference that assumes intrinsic betterness which will be the object of von Wright's logic.[8] Moreover, only where the above is the case can one say that the *intrinsic* sense of preference is involved.

According to von Wright, the relation which preference has to choice is less than tangible, in that one can entertain an intrinsic preference and yet not choose any deliberate course of action.[9] As for example, one may prefer a sunny day to a storm, without having any choice as to which state will be finally actualized on a given occasion. Therefore, one is in the position of defining in precise terms the area of logic of preference upon which von Wright endeavors

---

[6] Von Wright, Georg Henrik, *The Logic of Preference*, p. 10.

[7] Von Wright, Georg Henrik, "The Logic of Preference Reconsidered," p. 144.

[8] Von Wright, Georg Henrik, *The Logic of Preference*, pp. 14-15. Von Wright's differentiation between the intrinsic and extrinsic sense of preference rebuffs R.E. Jennings' criticism in "Preference and Choice as Local Correlates," *Mind*, (1967), vol. 76, pp. 556-558. On this point R.Z. Parks successfully defends von Wright against Jennings' charge of inconsistency, see Parks' "On Jennings on von Wright on Preference," *Mind*, (1971), vol.80, pp. 288-289.

[9] ibid., *The Logic of Preference*, p. 16.

to concentrate, namely the area which deals only with the intuited relations of preference in an intrinsic sense, at the exclusion of considerations of preferential choice, i.e., preference in an *extrinsic* sense.

At this juncture an important question can be raised as to the defensibility of the idea that one can "prefer" in an intrinsic sense, independently of actually choosing what is preferred in a spatiotemporal environment. Von Wright claims that intrinsic preference, so called, is *in ordinary discourse* allied to the notion of "liking," both forming the expression of a dispositional attitude needing no objective reference. Indeed, further on, it is just this kind of connection between preference and liking, which allows him to claim that his work and Soren Hallden's investigation into the "logic of the better" have a close relationship. For preference, conceived by von Wright as an expression of hedonic desire, is part of the *broader* concept of better, and the hedonically pleasing (i.e., the better) *is* the preferred. Proceeding along these lines, however, there is the problem of grasping with some measure of clarity the nature of the state of affairs which preference is said to convey in this "intrinsic" sense.

At the outset, von Wright claims that preference should be considered on a conceptual level within a context of anthropological and axiological ideas. This would ordinarily involve a sense of state of affairs which is basically extensional, in that reference is being made to persons, actions, occasions, segments of time, etc., which would be the constituents of states of affairs. Yet, though von Wright claims that preferences in his sense must involve the aforementioned elements, the sense of preference which he selects as basic to his inquiry is one which operates under conditions of *ceteris paribus*, and which avoids all reference to an extensional interpretation of a preference expression.

Thus, as expressed by preference, the state of affairs which constitutes the *preferring* by some agent with regard to some object is not an event of a spatiotemporal sort. Rather, von Wright appears to be suggesting here a kind of state of affairs which is more internal, in the manner of say someone having a good mental state, e.g., the liking of x over y, as a personal expression of a hedonic good. Hence, his claim — that he is dealing with states of affairs and that this is the "object" of his analysis — must be approached with the utmost of caution.

The issue here should be clarified in that one is not intending to confuse "liking" as a private mental state with one's object of preference. Rather, what is intended is to draw out the distinction between a mental state of affairs as a significantly different *kind* of occurrence than that of say the experimentally testable event of agent X preferring *a* over *b*, at time *t*, for reasons *x*, *y*, and *z*. It seems clear that what von Wright is saying is that preference in the intrinsic sense involves desires and likings as one's personal disposition toward certain objects, totally apart from considerations having to do with their extension. Moreover, one may add that these *preferred* objects, within the context of von Wright's analysis, have a unique phenomenological status, in that they are objects of one's desire, and not actual things in a spatial dimension. The *liking* of an object can surely be said to influence one's perception of it differently than where that same object is held to objective scrutiny. Thus, it can be said that, in essence, the kind of state of affairs which underlies von Wright's thinking at this point is one which is private and personal, though alluding to things as objects of desire. Von Wright's approach here is no doubt dictated by his overall view of a logic of preference as a formal system of valuation, being applicable to any specific situation precisely because it is not dependent upon any extensional context. Evidently, valuation is seen as the liking of some object, and the formalization of the discourse concerning this liking is the logic of preference.

Related to the above is the matter pertaining to the nature of the analysis which von Wright pursues throughout. Surely, he claims that he will be employing propositional logic as an investigatory tool so as to clarify the formal character of the preference-relations he intuits. As such, propositional logic does not impose any restrictions upon which propositions will be designated by variables 'p', 'q', 'r', etc., as long as the propositions themselves are in the assertive mood, and their truth value can therefore be determined. The only constraint propositional logic imposes is that its relational connectives be truth functional as specified within the logic itself. In the context of what von Wright is proposing, the variables p, q, and r must range over sentential expressions (propositions) about "liked" objects. Here it is interesting to observe how 'p' and 'q', in the simple preference expression 'pPq', range over expressions of occasions or things desired or not desired. Roughly, one appears to have here something like 'p' expressing the sentence: "The bringing about of x is intrinsically good.", and 'q' expressing the sentence: "The bringing about of y is intrinsically bad." The positioning of p before q, separated by 'P', for the relation "is preferred to", establishes the prioritized order required of an expression of preference between p and q. This, at the very least, is a rendering of the basic sentential relationships of von Wright's view on the structure of preferences found at the outset of his treatise.

Some points of tension may be found in his view, especially where one inquires into the ontological status of these preferred occasions, objects, events, etc. As noted earlier, these are *objects of desire* and thus not necessarily knowable as objectively real things. Their status as entities of some sort comes very close to what can be encountered in discussing Roderick M. Chisholm's analysis of the logic of the intrinsically better, with the clear influence on the latter by Meinong's theory regarding *Objekte*. However, in the absence of any claim by von Wright that he is working from such a phenomenological context, one can only raise the question at this point, and hope for clarification further on as to his conception of the nature of these objects of preference. The issue here is not superfluous, since it is intimately connected with the question of how one determines the truth value of the sentential *relata* which help constitute the truth value of the entire preference expression itself. Basically, how does one attribute truth value to a sentential expression concerning the intrinsic goodness of x, when x is not necessarily knowable through public means? This question is, of course, assuming that "intrinsic goodness" is an unambiguous notion, when in fact it is not.

It could be argued, — and this is a vitally important point —, that where one solely takes von Wright's position simply on the level of it being a logic of preference within context of *ordinary language*, where the depth of philosophical concern is minimal at best, then one can follow what he has to say as an interesting application of propositional logic.[10] Indeed, the latter position would seem to allow for continuing with an exposition of the strikingly ingenious things von Wright discovers concerning the formal relations among preferences.

Though the above attitude could be adopted for the purpose of getting on with an exposition of von Wright's work, the basically philosophical difficulties will invariably emerge in the course of that exposition, and they cannot be indefinitely deferred. Not only can we question whether the ordinary language context would permit the use of an expression such as "intrinsic goodness," but there is serious doubt as to the cogency of one preferring in an intrinsic mode, independently of any consideration of choice within a contingent spatiotemporal realm, i.e., under conditions of *ceteris paribus*. Even in the example cited by von Wright at the outset of his treatise, purporting to illustrate the possibility of intrinsic preference, it hardly seems defensible to say within the context of ordinary discourse that one prefers a sunny day to a stormy one,

---

[10] ibid., p. 15.

apart from a context of things to be done and actions to be taken. Apparently, von Wright is willing to allow for the possibility of an "idle preference", whereas preference inherently seems to convey some semblance of dynamic activity. Here Alan R. White's insightful observation is in order, for as he notes in *Modal Thinking*, the logic of discourse involving preferences requires a context of needs to lend it intelligibility. In essence, for White there seems to be an undeniable "end state" connected with preference.[11] Thus, at a basic conceptual level, there appears to be a formidable array of possible problems connected with the way in which von Wright wishes to define his terms, e.g., preference as context-free and yet as involving action.

Von Wright assumes a knowledge of the simplest operations of propositional logic. He introduces the idea that the variables 'p', 'q', 'r', ... etc. schematically represent sentences describing *generic* states, i.e., states that they may or may not obtain on any given occasion. Thus, P-expressions will be those of the form pPq and (p&q)Pr, which are read as the state p is preferred to the state q, and the state of p and q is preferred to the state of r. Von Wright informs the reader once more that he is considering preferences in terms of subjects and occasions, and that P-expressions of the kind introduced above are of this nature. Moreover, he goes on to claim that individual preference expressions can be combined by means of truth-functional connectives to form molecular compounds of more complex preference expressions. Illustrative of molecular P-expressions is the following formula: -(pPq) ⊃ (p&q)P-r, which is to be read as, if it is not the case that p is preferred to q, then it is the case that p *and* q is preferred to not r. This entire expression is taken relative to some unspecified subject and occasion. Moreover, it is helpful here to bear in mind the point he makes in his later essay that intrinsic preference must be irreflexive, since one cannot have pPp.[12]

Significantly, von Wright's allusion to "proposition-like entities" is made for the expressed purpose of using Boolean connectives to express P- relations. In correspondence with this author, von Wright finds no problem with this procedure, arguing that though the formal definitions of implication and negation are often fraught with difficulties, they are not anymore problematic here than elsewhere, irrespective of the fact that symbolic logic is being used to

---

[11] White, Alan R., *Modal Thinking*, Cornell University Press, Ithaca New York, 1975, p. 113

[12] Von Wright, Georg Henrik, *The Logic of Preference*, pp. 19-21. (see also Donald Davidson's "The individuation of Events," in *Essays on Actions and Events*, Clarendon Press, Oxford, 1980, pp. 168-167.)

formalize preference-relations. Rather, P-relations are said to be expressible in symbolic terms in virtue of the fact that the proposition-like entities over which 'p' and 'q' range are simply one of the many possible interpretations which 'p','q','r', etc. are said to have in Boole's rendition of algebraic logic. Surely, the importance of Boole's insight was in his observing that the definitions of logical connectives were distinct from any consideration regarding the content of what variables represent.

Apart from von Wright's remonstrance to the contrary, there is ground for questioning what proposition-like entities are, and how they are to be construed. No doubt von Wright considers this concept crucial to his analysis, since it is a conceptual point of departure from the approach taken by Hallden and Aqvist. However, the cogency of this concept, together with the mode of von Wright's method of operation is challenged by Aqvist in "Chisholm-Sosa Logics of Intrinsic Betterness and Value." Aqvist recognizes the dubious character of 'P', as it is said to associate with truth-functional connectives. Moreover, Aqvist also doubts the wisdom of maintaining that goodness, badness, indifference, neutrality, etc. can be expressed meaningfully within a framework of propositional logic. As was seen, his own solution would be to introduce a modal logic which would sharpen the scope of these concepts to a more manageable level. As things stand, Aqvist finds von Wright's perception of intrinsic preference-relations much too unyielding to specificity, at least to point of being expressed formally.

Examining the *rationale* for von Wright's introduction of the idea that his logic of preference will deal with sentential expressions having reference to "generic" events is vitally important. Essentially, this is his way of underscoring the thesis that his logic will handle the situations and states of affair preferences involve on an impersonal and atemporal basis. However, von Wright's assumption concerning the role of such events has not escaped the incisive criticism of Donald Davidson. The latter observes that there is really no basis on which to claim that a sentence in everyday discourse refers exclusively to a generic action, as opposed to its referring specifically to particular action. For example, von Wright argues that the sentence 'Brutus killed Caesar' is ordinarily taken to be about the unique and specific event of the killing of Caesar by Brutus, whereas the sentence 'Brutus kissed Caesar' is ordinarily taken to be about a generic action referring to a kind of event, i.e., the repeatable occurrence of Brutus kissing Caesar. However, it can also be readily illustrated that *though* the sentence 'Brutus killed Caesar' is usually rendered as being about the specific event belonging to the genus of "a killing of Caesar by Brutus," alternatively the sentence 'Brutus kissed Caesar' can also be rendered or paraphrased as being about the particular event, belonging to the genus of a "kissing of Caesar by Brutus" event. The important point here is that there appear to be no absolutely necessary reasons for insisting that certain sentences must be about certain events, whose nature is said to be generic. This insight by Davidson, though employed in attacking von Wright for a view expressed in *Norm and Action*, has highly significant consequences for the latter's work on the logic of preference. Surely, what Davidson says here on how sentences refer to different types of events cuts deeply into von Wright's acceptance of the position that on the level of ordinary discourse, there are sentences which unambiguously refer to events which are impersonal and atemporal, i.e., generic events.

At this point one should review the apparent intent of what von Wright is proposing, namely a formal study of preferences in relation to potentially specifiable states of affairs. Interestingly, he points out that 'p', 'q', etc., will stand for *sentences* which express states of affairs concerning preferences. Thus, the atomic and molecular expressions he brings out above are wholly and completely sentential expressions. His aim here is, of course to be consistent with his adoption of the system of propositional logic as a tool by which to present his logic of preference. Thus, he must deal with sentences which express propositions so as to have access to operations within propositional logic. However, his insistence — that the sentences which express the states of affairs of what is or is not preferred are to be taken in an intentional sense, i.e., they are about unspecified subjects and occasions, — renders the use of the implicational sign ' → ' difficult to comprehend.

Surely, von Wright wishes to differentiate between actual or specified subjects and occasions, and unspecifiable subjects and occasions, i.e., "as atemporal and impersonal..."[13] However, it is with the latter that the problem seems to be most persistent. What is an unspecified subject and/or occasion? The difficulty here is precisely at the point of saying that 'pPq' is significant, but 'p' and 'q' are unspecified. What then is the significance of 'P' within the logic? This question should bring home the point that the meaning of 'P' is taken not in a sense that emerges within the proposed logic von Wright sets out to present. Rather, it derives its sense as a primitive which we are to adopt along with the basic tools of the propositional logic itself. Thus, 'P' in and of itself does not explain anything, in that some other symbol could have been

---

[13] Von Wright, Georg Henrik, "The Logic of Preference Reconsidered," p. 144.

used just as well to stand for the relation of preferring. Apart from the dispensability of 'P' is the question of what sense it makes to say for example, that 'pPq' expresses how state of affairs p is preferable to state of affairs q, where 'p' and 'q' do not specify a particular state of affairs. Does it make sense to say that one can prefer within the context of an unspecified state of affairs? Furthermore, the difficulty here becomes even more curious in light of the fact that von Wright wants to consider preference only in an intrinsic sense, which he proceeds to explain in terms of the "liking" for one state of affairs as opposed to some other. Thus, 'P' is to be construed as some sort of liking for an unspecified state of affairs p in contrast to some other state of affairs q. Here again one is unsure how to make sense of saying that one prefers (or likes) an unspecified state of affairs as opposed to some other state of affairs.

Perhaps the above can be allayed by saying that von Wright is actually concerned solely with the "form" of expressions dealing with preferences, and that this alone is the direction from which his logic should be approached. On this score it seems equally reasonable to ask whether the form, so-called, is reflective of the application of propositional logic itself, or is it somehow peculiar to the character of preference expressions? Here again one faces the problem of how to interpret the implication sign in von Wright's logic. Implication is not a subjective relation. Yet, von Wright's insistence that 'p' and 'q' must stand for unspecified subjects and occasions of liking makes the sense of the implicative relation very unfamiliar in contrast to its normal use in propositional logic. Surely, it is to be recalled that he considers 'p' and 'q' as schematic representations of sentences expressing generic states. Yet even here it may be asked of what is there a schematism? Is it a schematism of the sentences expressing the generic state of affairs, or is this a schematism of the generic events or states themselves? Von Wright would most likely claim that it is a schematism of the sentences themselves, rather than of the generic states they refer to. As such, however, a schematism of sentences describing a generic event does not provide much by way of insight into the formal structure of any particular state dealt with by a sentence, which in this case is the mental state of preferring something.

The bottom line of the criticism above is to refute von Wright's basic assumption that one can have a logic of preference, without needing to allude to a specific or particular state of affairs. In other words, it is dubious that one can have a logic of preference whose objects are propositions dealing exclusively with unspecified state of affairs, e.g., generic events. This very thesis is disputed by Alan White, who questions the intelligibility of claiming that modals can qualify either sentences or propositions. While not directly referring to von Wright's work on preferences, White takes to task von Wright's pervasive acceptance of the view that modals do qualify propositions, whereas in White's view modals can only be considered as signifying relations between things. The distinction White brings out is that of a subject of predication being considered *de dicto*, and one being considered *de re*. According to White, it is vitally important that one realize that "... modal concepts do not signify particular items either *in the world* or in *our minds*, but *the relation of one item to others in a situation.*" (italics added) White argues further that one simply cannot have modality *de dicto*, since modals must address the manner by which things are related and not the manner by which things are spoken about. What is preferable is not, for example, the proposition *that* X is Y, but *what is expressed* or *stated by* that proposition. Even where one has the cases of saying: 'it is possible that such-and-such is true', or 'that such-and-such a proposition is perhaps true', a proposition itself is not being signified, but rather it is the proposition's *being true* which is signified by the use of the term 'possible'. White cautions as to the danger of assuming that there can be a modality *de dicto*, since it leads to the adoption of modal properties, such as the *being* possible, the *being* necessary, and in the case in point one may say the *being* preferable. Properties such as these are often erroneously construed as characterizing thought, completely disregarding the essential point that modals can only be said to characterize objects of thought, i.e., things or that which is thought about. Surely, von Wright's concentration upon intrinsic preference is aimed to focus upon an attitude of thought which is directed toward sentences about what is *intuited* as intrinsically good, bad or indifferent; within an artificial context of *ceteris paribus*. For this reason, White's position challenges the very likelihood of having a logic of preference, which is about things, i.e., the extensionally orientated.[14]

Apart from these preliminary difficulties, it is interesting to follow further the unfolding of von Wright's logic, as a way of understanding the significance of his effort. Von Wright does not see his formal system as an axiomatized system, rather he considers it as a technique with which to determine whether or not a sentence expressing a preference is formally true. As he stated at the outset of his inquiry, his logic is to be taken as a system of elementary valuation. Towards this end he presents five principles relating to preferential expressions, two of which fall into one group, and the other three fall into a second grouping.

---

[14] White, Alan R., *Modal Thinking*, pp. 171-172.

In the first case, one has principles relating to the formal properties of the relation of preference. The second case has principles of transformation, whereby P-expressions can be transformed into a standard form, to which the decision technique made possible by principles in the first grouping can be applied.

The first two principles in the former grouping deal with the asymmetry and transitivity of the preference-relation. Ostensibly one can say that the relation of preference is asymmetric if state a is preferred to state b means that state b is not preferred to state a, for the same subject and occasion. Also, roughly the transitivity of the preference-relation is where a is preferred to b, and b is preferred to c, for the same subject and on the same occasion, aPc.[15]

Having stated the above, von Wright is sensitive to the fact that there is certainly a problem with his rendition of the transitive sense of the preference-relation. For it could very well be the case that a person may prefer apples to oranges, and oranges to peaches, but it need not follow that on any given occasion he will prefer apples to peaches. Evidently, where one is consistent in maintaining an intrinsic sense of preference, there is invariably a conflict with the practical realities of idiosyncratic taste. Hence, even at the outset von Wright recognizes the incessant problem the enterprise on which he is embarking upon poises, namely the difficulty of fitting the "logic" of preference on to the actual practice of preference. He ends this section of the discussion by simply *assuming* that preference is a transitive relation, though recognizing that there are problems connected with this assumption.[16]

Von Wright asks, "What *is* it (what does it mean?) to prefer one state of affairs to another?" Recognizing that the complexity of this question makes it virtually impossible to take into account all the ramifications of this question, he limits himself to a scrutiny of the possibilities resulting from *saying* that one prefers generic state p to generic state q. On a strictly formal basis he sees four possibilities in which p and q can be related, namely, world: p&q, world: p&-q, world: -p&q, or world: -p&-q. In the situation where p is preferred to q, one can say that the person prefers retaining p and losing q, hence he would rather have p&q. Thus, if his world contained both p&q, he would rather see it change to p&-q, rather than to -p&q. Also, if his world was that of -p&q, he would rather see it change to a world where p&-q were the case, rather than to have it remain the way it is. In the same vein, if the person's world were that of -p&-q, and if he were to prefer p to q, it would mean that he would rather see his world change from what it is to one of p&-q, or to achieve p at the expense of losing q.

---

[15] Von Wright, Georg Henrik, *The Logic of Preference*, p. 21.
[16] ibid., p. 22.

In summary, saying that p is preferred to q is equivalent to claiming that the subject prefers end-state: p&q to that of -p&q, '*as end states of contemplated possible changes in his present situation* (whatever that be)'.[17] This constitutes von Wright's third rule, and it is included in his second grouping of principles governing his logic.

Thus far, it is important to observe the manner by which von Wright connects preferences with "contemplated" possible changes in the subject's present state. This is not a reference to actual spatiotemporal changes, but to possible and contemplated change. Furthermore, 'p' and 'q' were found to schematize sentences about possible state of affairs. Hence, von Wright is making a very concentrated effort to steer away from any spatiotemporal meaning for state of affairs. He endeavors to fit his formal analysis of preferences within a framework of possible states, wherein the logical consequences of expressions of preference are considered quite independently of particular actions. Presumably, the allusion to "contemplated changes" is directed toward the idea that one has here the rational assessment of rational alternatives with regard to possible-world states.

The unfolding of von Wright's logic needs scrutiny precisely at the point where he assumes that to prefer whatever is expressed by p means that therefore one also prefers -q, where p is preferred to q. It is the implicit assumption of the negation of q which seems troublesome, or at least in need of further clarification. Surely here one is dealing with an inference which is based upon the asymmetry of p being preferred to q. Yet, it becomes apparent from von Wright's own discussion of the assumed transitivity of 'P' that the preference of x to y need not necessarily preclude the possibility on some other occasion of y being sometimes preferred to x, by the same person. Hence, here it is the sense of the word "necessarily" which brings the issue of the relevance of von Wright's logic to a sharp focus.

The above cannot be a logical necessity, since von Wright is setting up his analysis in terms of the "possible" world model. The necessity here would presumably be meant in the modal sense of not *necessarily possible*, and yet this would not justify interpreting the function of 'P' as that of "negating" the possibility of y in the expression 'xPy'. In a sense it could be said that within the above context it is inconceivable that a state of affairs itself can be negated. Rather, it seems more plausible to argue that what is negated is the possibility or nonpossibility of a proposition expressing a state of affairs. This is to say

---

[17] ibid., p. 25.

that whereas it makes sense to negate sentences such as may be expressed by 'p' and 'q', for von Wright it makes no sense to say that one can negate possible-world states expressed by such sentences. However, there is something amiss in trying to equate the negation of a proposition with the impossibility of a world state itself. The former concept deals more with the manner in which a content is expressed sententially, whereas "impossibility" evidently should be attached to the content itself. This again seems to be another manifestation of the distinction of predication *de dicto* and predication *de re* mentioned above.

Von Wright's fourth principle attempts to explain the nature of the distributive character of preferences dealing with disjunction. The principle he presents is carefully dissociated from any connotation involving probability and risk. Thus, it is to be seen entirely within the context of possible-world states. The principle which follows is expressive of disjunctive preference, i.e., (pvq)Pr, and it states quite simply that disjunctive preferences are conjunctively distributive.[18] The rule initiates for von Wright the need to differentiate between preference as considered without the factor of *risk*, and preference considered from the angle of probability where *risk* is a factor. The latter is usually what is treated in economics, and von Wright recognizes the fact that in this latter sense preference cannot be subjected to formal characterization.

The above is illustrated by von Wright as follows. In a case where a person is offered the following option: an increase in pay or longer holidays *to* fewer working hours in the day. Assuming that the increase in pay and the longer holidays are not exclusive of each other, then one can surmise that the individual may prefer (or like better) an increase in salary and longer holidays but no reduction in his hours of work. Alternatively, it could be said that he may prefer an increase in salary but neither longer holidays nor a shorter work week. Or finally, he could be said to prefer longer holidays to an increase in pay and a shorter work week. Thus, the disjunctive preference could be reduced to a series of conjunctive statements covering an increase in pay and longer holidays; an increase in pay and not longer holidays and a shorter work week: and longer holidays with no increase in pay and no shorter work week. This is what von Wright means when he says that disjunctive preferences are conjunctively distributed.

The matter, however, acquires a further complexity which prevents the above distribution where risk is allowed to enter as a consideration, for then there is only a probability that the subject will receive an increase in pay if he

---

[18] ibid., p. 26.

were to prefer the first option as opposed to the second or third, or both the second and third, etc. Under these new conditions, which again are seen by von Wright to be at the foundation of economic theories concerning the preferred, the uncertainty as to what will actually happen creates the situation where even if one were to choose the most cherished option — receiving the increase in salary — there is the likelihood that he will still receive the shorter work week. Consequently, where all of the options within the range of preference are only of a certain probability as to the likelihood of their occurring, one cannot consider these options as in any way conjoined. For preference involving risk is such that one is not sure *prior* to the choice that what he chooses will turn out to be the case. Consequently, one is not sure whether what he chooses will be what will actually come to pass, thus all acts of preferring in such a context must be considered separately.

Von Wright's view on the conjunctive distribution of disjunctive preferences did not go unchallenged by R.E. Jennings, in his note entitled "Or". Concentrating on von Wright's use of 'v' in '$(pvq)Pr \rightarrow (pPr)\&(qPr)$', Jennings argues that "... whereas for some relational expressions, adjacent 'either. .or' expressions are usually disjunctive, for other relational expressions, adjacent 'either ... or' expressions are usually conjunctive." The insight Jennings derives from his observation of spoken discourse, and *not* from the truth or falsity conditions governing disjunction in logic. Thus, in the case of 'I prefer either ice cream or pudding to cake' he says that one is not dealing with a disjunctive preference *at all*, in that in this case the 'is preferred to' expression is conjunctively distributive over the particular use of the word 'or'. That there can be cases where the 'either ... or' expression is taken conjunctively but in a *noncombinative* sense is illustrated by the example where one says 'Mary is heavier than either Jack or Bob', meaning that Mary is heavier than Jack, *and* Mary is heavier than Bob, but not that Mary is heavier than *both* Jack and Bob. The point here is that it is the context of linguistic use which determines the sense of the 'or' expression, and not the rules for the logical connective 'v'. The decision as to whether 'or' is to be taken disjunctively or conjunctively in its distributive role depends directly upon interpretative determinations made *prior* to formulating any preference postulates. In Jennings' view, von Wright is missing the vital function of linguistic usage, in its controlling influence over one's very perception of relations involving preferences.[19]

---

[19] Jennings, R.E., "Or", *Analysis*, vol. 26, (1966), pp. 181-183.

Historically, it is interesting to observe how von Wright, in his elimination of the role of risk from his approach to preference, endeavors to distance himself from the track taken by investigators such as Ramsey, Davidson, and Bohnert. For example, though von Wright expresses respect for Ramsey's remarkable achievement in this area, he pursues a line of inquiry which is not subject to the implications resulting from the application of probability, as is the case with Ramsey's inquiry. Von Wright goes to the opposite extreme, endeavoring to evolve a logic of preference within a context of all things being equal, a kind of frozen state of reality, where chance has absolutely no role in its influence upon human decision-making.

Apparently, the possibility of the formalization of preference — where preference is taken in terms of its features in extension — is rejected by von Wright as beyond any known range of productive inquiry. In sharp contrast, R. M. Martin's work which appeared in the same year as von Wright's, and it illustrated with marked clarity how an extensional logic of preference could be designed, with few of the debilitating drawbacks feared by von Wright. Of course, Martin's approach emphasizes the view that there is no one single logic of preference. Rather, there is the logical formalization of various types of preferential states of affairs considered in terms of their extension, with a language user, an experimental situation, and segments of time. On the other hand, von Wright's intentional approach builds upon a fundamental assumption; namely, that there is *a* fundamental logic of preference which can in some way be discovered. Indeed, his discussion of the way in which the logic of preference concerning things must ultimately assume the more basic logic of intrinsic preference involving propositions reflects the belief, in 1963 at least, that the logic of preference he is presenting is the most essential form of such a logic. However, von Wright does not attempt to justify what his intuition tells him on this score.

Von Wright's fifth and final principle deals with his conception of the *holistic* nature of preference.[20] Of the three separate ways in which he discusses the holistic character of preference, the most prominent of all will be that which involves "unconditional intrinsic preference," *ceteris paribus*. This is preference in the sense that, where all things are equal with regard to any total world state, a subject would prefer a change to p&-q rather than a change to -p&q, under all possible conditions. He cautions that on each occasion where he will refer to preferences without any further qualification, he will mean

---

[20] Von Wright, Georg Henrik, "The Logic of Preference Reconsidered," p. 145.

preferences only in the above sense. Thus, *The Logic of Preference* is devoted to the logic of "unconditional intrinsic preferences between states of affairs."[21] This view remains unchanged in his later article, "The Logic of Preference Reconsidered," since almost at the very outset he says, "A preference between states under specific circumstances or *ceteris paribus* will be called *holistic*."[22]

So as to avoid questions concerning the tenability of the notion of "unconditional preferences," von Wright stipulates that he will be positing a "Universe of Discourse," *relative* to which preference is said to be unconditional. This move is designed to eliminate objections as to how one can speak of preferential relations within a domain of an infinite number of possible states. Furthermore, it is also designed to by-pass objections dealing with unconditional preferences containing elementary states of affairs which are not in themselves truth functions of other states of affairs.[23]

In his review of von Wright's *Logic of Preference*, however, Donald Davidson carefully assesses the notion of "unconditional preference" as a basic presupposition of this logic. Davidson questions the intelligibility of the idea that preference can be unconditioned, thus having absolutely no affection by the actual state of the world. In this connection he also notes that one can point to no clear idea of what it would be like for our world to change in only one specific respect, "*and in no other.*" Yet, this is exactly what the condition of *ceteris paribus* demands. Von Wright's attempt to explain this notion does not fare well under Davidson's scrutiny, since it simply cannot be claimed by von Wright that the meaning of p is unconditionally preferred to q, where any other state of affairs, say r, has no influence on the preference in question, i.e., $(pPq) \rightarrow (p\&r)P(q\&r)$. What results from this sort of an explanation is a situation where normal formal inference breaks down. Surely it can be readily seen that though the preceding formalization is a logical truth, the consequent cannot be inferred from the antecedent, nor can one infer $(pPq)\&(qPr)$ from 'pPq' and 'qPr'. Davidson brings these points forward so as to question whether von Wright is really dealing with *logical truths* in his analysis, and whether any semantical interpretation can help to make clear the sense of a logic of preference, conceived under the condition of all things being equal. The general thrust of Davidson's remarks, therefore, is to argue that if a logic of preference is to have a distinctive character, one which is not simply the token manipulation of symbols, and

---

[21] Von Wright, Georg Henrik, *The Logic of Preference*, pp. 29-33.
[22] Von Wright, Georg Henrik, "The Logic of Preference Reconsidered," p. 147.
[23] Von Wright, Georg Henrik, *The Logic of Preference*, pp. 32-33.

tion of symbols, and which is to stand as a logic with clearly determined rigor, then it must have a sharper focus at least as to what the sentential connective 'P' means. Without this it virtually lacks comprehensibility. In summary, it may be said that Davidson is unwilling to allow von Wright's "Universe of Discourse," so-called, since it leads to a meaningless rendition of the formalization of preferences.

Within the above context, von Wright goes on to claim that "good" and "bad" can be defined very easily as p is good if p is unconditionally preferred to -p, and p is bad if its absence is preferred to its presence, i.e., -pPp. The sense of both these terms, i.e., "good" and "bad," is taken in an intrinsic as opposed to an extrinsic or instrumental sense. Thus, for example, good here is taken apart from any "consequences" having extrinsic value. In essence, von Wright observes that unconditional intrinsic preference is tantamount to a *hedonistic liking*, and it is not at all to be construed in terms of *moral approbation*.[24] The latter would invariably bring into play considerations involving consequences to one's actions.

Von Wright accepts as axiomatic the idea that P-expressions can be transformed in a manner which is logically tautologous, without in any way altering the truth value of the original expression "... provided that new variables are not introduced in these expressions through the transformations ..."[25] Consequently, he proceeds to show how three fundamentally logical operations — namely conjunction, distribution, and amplification — are readily applicable to P-expressions, within a prescribed Universe of Discourse.

(a) *Conjunction*

Given the atomic-constituents of the P-expression (p&q)P-r, the operation of conjunction allows that one can conjoin the expression to the right of P on to the expression to the left of P, if the former is negated, and the same can be done with the expression to the left of P, without there being any change in the original value of the expression. Thus, with respect to the expression symbolized above one has (p&q& — r) P(-r&-(p&q)). Moreover, using the disjunctive normal forms for the atomic-expressions constituting the elements of the above preference expression, it can be transformed as:

(p&q&r)P(-p&q&-r v p&-q&-r v -p&-q&-r).[26]

---

[24] ibid., p. 35.
[25] ibid., p. 36.
[26] ibid., p. 37.

(b) *Distribution*

In this operation, one simply replaces the atomic P-expressions in an expression such as the one derived above through the operation of conjunction, according to the rule of conjunctive distribution. Hence, the above becomes:

[(p&q&r)P(-p&q&-r)] & [(p&q&r)P(p&-q&-r)] & [(p&q&r)P(-p&-q&-r)][27]

(c) *Simplification*

This operation takes into account the variable s, which is interpreted as in the Universe of Discourse of the P-expression, but it does not appear in the original P-expression. The variable s will occur in some conjunctive expression of the P-expression somewhere along the line. This is handled by von Wright in a way where s is attached by conjunction to the P-expression in both of its two possible truth value variations. Hence again, given the above P-expression (p&q&r)P(p&-q&-r), one has:

[(p&q&r&s)P(p&-q&-r&s)] & [(p&q&r&-s)P(p&-q&-r&-s)].[28]

Finally, within this phase of his discussion, von Wright defines a "state description" within a Universe of Discourse as a "conjunction of all the variables and/or their negations."

Within the context of the above, von Wright proceeds to present one of his most significant insights into the formalization of preference, i.e., the idea of *"preference-tautology."* By way of illustration he presents the outcome of the operations of conjunction, distribution and amplification upon the following formula (pP-p)&(-qPp) → (pPq).[29]

Employing conjunction one finds the following result:

(pP-p)&(-qPp) → (p&-q)P(-p&q).

With distribution here being vacuous, von Wright proceeds to apply the operation of amplification to the above:

[(p&q)P(-p&q)] & [(p&-q)P(-p&-q)] & [(p&-q)P(p&q)] & [(-P&-q)P(-p&q)] → (p&-q)P(-p&q)

Here it is seen that each of the four preferential conjuncts preceding the implication (p&-q)P(-p&q), and including the latter, is such that the negation of state of affairs p or q, but not both, occurs either to the left or to the right of

---

[27] ibid., p. 37.
[28] ibid., p. 38.
[29] ibid., pp. 41-42.

each P-relation. Von Wright observes that the asymmetry of preference subscribed to at the outset of his inquiry does not limit the distribution of truth values for the above five conjunctive P-expressions. The only limitation results form the transitivity of the P-relation, where if '(p&-q)P(-p&-q)' and '(-p&-q)P(-p&q)' are true, then '(p&-q)P(-p&q)' must be true as well. According to von Wright, it is this last fact of transitivity which makes the expression derived from amplification a "preference-tautology". Clearly, there is here a strong *logical* bond between the antecedent and the consequent of the proposed implication.

At this point, von Wright wishes to have the reader recall that 'good' and 'bad' were defined earlier in terms of what is or is not preferred. Thus, he claims that the formula "proved" above can now be read as "A good state of affairs is to be preferred to a bad state of affairs" or more simply "Good is better than bad."

Von Wright does not see the suggested interpretation of the formula above as a surprising result, but it is nonetheless a truth established by his logic of preference.[30] He claims that the above is a *logical* truth, but adds immediately that it is not a truth of "ordinary" logic. Also, such truths are not readily seen by him to be truths resulting from merely the stipulation of a definition and are hence not to be seen as trivialities.

Furthermore, in what is perhaps the most crucial point of the entire book, von Wright says that his logic is *not completely* extensional with the logic of propositions, where new variables are allowed introduction. Rather, the logic of preference as conceived by him can be said to be only *semi-extensional* with the logic of propositions, since the introduction of variables which do not occur in the original P-expressions are not permitted. This provision is designed to preserve the conception of preference which pervades his entire treatise, namely that of unconditional preference in an intrinsic sense, *ceteris paribus*. He argues that if his logic were to allow for the introduction of new variables then the "holistic" sense of preference with which he is working would be untenable.

With respect to his observation that the logic of preference he envisages is expressive of logical truths, though not in the *ordinary* sense of logical truth, it is important to bear in mind that von Wright does not evolve his logic as a deontic logic.[31] Nowhere does one find von Wright saying that p "ought to be"

---

[30] ibid., p. 42.
[31] Von Wright, Georg Henrik, "The Logic of Preference Reconsidered," p. 145.

preferred to q, or that it is obligatory that p be good, and q bad. In his later essay, von Wright makes it quite clear that the logic of preference should be distinguished from the logic of preferability. It is in the latter that one must consider the formal properties of locutions dealing with "one ought to prefer x to y." However, the overall aim of *The Logic of Preference* is to develop a logic or formal theory of value in general, using as much of propositional logic as possible.

In "The Logic of Preference Reconsidered" von Wright comes to doubt any claim of success in achieving the aim discussed above. For he finds that in attempting to equate the good with the preferred, certain serious and potentially destructive consequences emerge. First, he realizes that if one were to argue in the manner suggested in his earlier work, then under the demanding sense of preference which involves total states of affairs in the *ceteris paribus* mode, there can be only two states, one good and the other bad. In other words, in the "holistic" sense of preference which pervades his early work, there cannot be any good or bad states *between* what is preferred and what is not preferred. Rather, there can be only value-neutral states between these two extremes. Secondly, where preferences are considered in the context of saying that only on certain occasions is, say, p preferred to q, then the states which lie between p and q may be ranked as good or bad or as better or worse relative to their opposites. However, it works out in such a way that all the better states would be equal to each other relative to their value, and all the worse or bad states would be of equal value. But there would be no degrees of good, and of course, there would be no degrees of bad. For von Wright, these two results violate certain intuitions which he finds basic, as for example: "... *good things may be better or worse among themselves, and similarly bad things*. ..."[32] Ironically, his possible-world analysis of preference-relations is found by him to be incapable of providing for a preference-ranking, and as it stands his logic of preference in *The Logic of Preference* is suitable only for a non-relativist theory of ethics where there is sharp dichotomy between what is good and what is bad. In light of von Wright's own observations on these points, it is now dubious whether his objective of achieving a "general" value theory has been, or rather can be achieved. Further, in the later essay von Wright endeavors to present a logic of preference, which while preserving the intrinsic sense of the term, admits of preference-rankings. However, apart from the merit of this effort, which will be assessed further on, his work in *The Logic of Preference* raises the

---

[32] ibid., pp. 160-161.

serious question of whether or not the intrinsic sense of preference can ever allow for a preference-ranking. This is to say that theoretically, if preferences are going to be taken in the sense of expressing a "personal liking," as claimed at the start of *The Logic of Preference*, then as a purely subjective state such liking cannot be said to allow for degrees. The liking of something, under von Wright's explicit condition of *ceterls paribus* is, as noted, not accessible to an objective standard, which would enable one to distinguish one case of liking as being of higher rank than another. This highly prohibitive restraint, i.e., contending with a private state which must be faced if von Wright's thesis is to be pursued in its own terms — apparently retards any likelihood of his evolving a formal value theory which can have wide-ranging applicability, at least with respect to saying in extensional terms that one thing is verifiably better or worse than another.

Yet, apart from the conceptual difficulties which seem built into von Wright's logic, he places great emphasis on what he takes to be the important fact that the formal relations of preference he intuits can be transformed into logical tautologies, reflecting the generally recognized requirements of propositional logic. The semi-extensional status of his logic with respect to propositional logic is not seen to be a shortcoming in claiming that one can have logical tautologies within a logic of preference.

Consequently, quite simply pP--q is the same as pPq, similarly (p&q)P-r can be transformed into the formula (q&p)P-r.

Thus also, (pP--q) ↔ (pPq) is a "P-tautology", since by applying the operation of conjunction to it one has: (p&-q)P(-p&q) ↔ (p&-q)P(-p&q), which is self-evidently true, or tautologous.[33]

Von Wright also illustrates the possibility of proving P-tautologies by replacing the expressions to the left and right of the P-relation with tautologically equivalent expressions whose disjunctive normal forms again manifest P-tautologies. As in the previous case, the stipulation is that the tautologically equivalent P-expression *"contains only variables, which occur in the given expression."*

His illustration of how tautological equivalence fails — where a new variable is introduced which is not in the original P-expression — is presented in terms of the simple formula pPq being replaced by: (p&r v p&-r) P(q&r v q&-r), and by conjunction one has (p&-q&r v p&-q&-r)P(-p&q&r v - p&q&-r). However, distribution yields the following set of conjunctive P-expressions:

---

[33] Von Wright, Georg Henrik, *The Logic of Preference*, pp. 43-44.

$$[(p\&-q\&r)P(-p\&q\&r)]$$
$$\&\,[(p\&-q\&r)P(-p\&q\&-r)]$$
$$\&\,[(p\&-q\&-r)P(-p\&q\&r)]$$
$$\&\,[(p\&-q\&-r)P(-p\&q\&-r)]$$

The very last conjunctive P-expression is not equivalent to pPq, since (p&-q&-r)P(-p&q&-r) is equivalent to (p&r)P(q&-r), which contradicts his fundamental notion of *unconditional* preference, *ceteris paribus*, through the introduction of r.[34]

Von Wright's inventiveness extends to an interesting discussion dealing with a diagrammatic method for testing the *consistency* of P-expressions. The approach is to set forth a square, so that at each corner there is one possible permutation of the 'p&q' conjunction, thus:

$$\begin{array}{ccc} p\&q & \longleftarrow & p\&-q \\ \downarrow & \swarrow & \downarrow \\ -p\&q & \longleftarrow & -p\&-q \end{array}$$

The restrictions for connecting the four points of the square are: (1) an arrow cannot lead in two opposite directions between the same two points, owing to the asymmetric nature of the P-relation; and (2) there are no pairs of points expressing negated constituents of the conjunctive pair p&q, which constitute the state of affairs represented to the left and to the right of the sign P, and between which an arrow runs.[35]

Upon determining that a P-expression is inconsistent, it follows, according to von Wright, that the negation of this expression must be tautologous.

As an illustration of how his diagrammatic method functions to determine the consistency of P-expressions, von Wright considers the following expression: (pP-p)&(-qPp)&-(pPq), which is transformed into "a molecular compound of P-constituents" as follows:

$$[(p\&-q)P(-p\&q)]$$
$$\&\,[(p\&-q)P(-p\&-q)]$$
$$\&\,[(p\&-q)P(p\&q)]$$
$$\&\,[(-p\&-q)P(-p\&q)]$$
$$\&-[(p\&-q)P(-p\&q)]$$

---

[34] ibid., pp. 45-46
[35] ibid., p. 49.

Diagrammatically, these P-constituents appear as follows in von Wright's presentation:

$$\begin{array}{ccc} p\&q & \leftarrow & p\&\text{-}q \\ \downarrow & \swarrow & \downarrow \\ \text{-}p\&q & \leftarrow & \text{-}p\&\text{-}q \end{array}$$

The dotted arrow clearly demonstrates the relation of transitivity holding between p&-q and -p&q, due to the preference-relation between p&-q and both p&q and -p&q. More significantly, perhaps, the diagram also shows the inconsistency of the formula under investigation, since the diagramming of the dotted line, i.e., preference expression (p&-q)P(-p&q), is in clear violation of the fifth molecular constituent brought forth above, which states that there cannot be a preference-relation between p&-q and -p&q. Having thus shown the inconsistency of (pP-q)&(-qPq)&-(pPq), it follows that its negation must yield a tautology.[36]

Von Wright proceeds to illustrate the inconsistency of a few *negated* "(meta) theorems." The cases in point deal with the two expressions: (1) (pP-p)&(qP-q)&(pPq), which is interpreted as saying "Of two good states of affairs, one may be better (worse) than the other." and (2) (-pPp)&(-qPq)&(pPq), read as saying: "Of two bad states of affairs, one may be worse (better) than the other."[37]

Apart from the demonstration of their consistency with his method of diagramming, the two meta-theorems presented above are interesting for additional reasons. First to be observed is von Wright's introducing into the analysis of preference in a more pronounced fashion than ever before the verbal interpretation of his insights. In so doing, he is utilizing his earlier claim that the notion of better is somehow more all-inclusive, and that the logic of preference in a way is "contained" within the formalization of the notion of the better.[38] Thus in (1) and (2) above, one sees how the understanding of certain relations of preference leads on to claim that a certain state may be better than another because it is preferred, and in being preferred it must be good. Here a three-level conceptual stratification appears to be operating, namely first the

---

[37] ibid., p. 50.
[38] ibid., p. 15.

basic level of that which is unconditionally preferred, secondly the level of good and bad relative to what is or is not preferred, and finally the higher level of what is better or worse, relative to what is good or bad.[39]

However, his interpretation of (1) and (2) offers virtually no insight into von Wright's conception of good and bad in terms of negated and non-negated states of affairs. For example, very little is afforded by way of explication of the relation between negated states and the bad. And it is tempting to accept as mere verbal stipulation what he has to say here. However, as noted earlier, he is strongly opposed to the conclusion that what he has to say about the meaning of 'good' and 'bad' is merely trivial definition. Thus, one wonders why it would not be permissible to have a negated state of affairs as descriptive of something which is good. Surely, not wanting a nuclear war is generally recognized as a good or, perhaps more basically, a "preferred" state of desire to see come about. Yet, is it to be said that, because this is a negated state of affairs and is expressed in terms which involve negation, that it is bad? Von Wright's reply here is that it is simply not that the negated states of p and q *alone* are equated with the bad, but rather that which is "not preferred" is said to be bad, whereas that which is preferred is said to be good. Hence, p is bad if one were to prefer: -pPp. Consequently, it is the preferring *in connection* with the preferred state which determines the goodness or the badness of what is preferred. However, quite apart from his great strides in this area of prohairetic analysis, one still encounters an unexplained tendency in von Wright to read into the negation sign meanings which it does not have in propositional logic. For example, in comparing the meta-theorems (1) and (2) above, one finds that symbolically their only point of differentiation is in the fact that the use of the negation sign in (2) is absent in (1), e.g.: (1) (pP-p)&(qP-q)&(pPq) and (2) (-pPp)&(-qPq)&(pPq). Surely, a reading of (1) and (2) in terms of good states and bad states could be supported only by attaching to the negation signs as expressed in (2) a signification not ordinarily found in propositional logic.

Furthermore, von Wright's analysis becomes "curiouser and curiouser" since he goes on to speak in terms of it being better that both of two states of affairs be *present*, than that only one be present, or that both be *absent*. The suggestion here is that not only do negated states convey "bad" states, -p, (together with the act of preferring), but they are to be interpreted as somehow intimating the *absence* of that world state, say p. This goes far beyond the fundamental assumption of propositional logic that the negation of an expression

---

[39] Von Wright, Georg Henrik, "The Logic of Preference Reconsidered," pp. 160-1

is no less an assertive statement of the logic than its non-negation. Basically, it is his blending of propositional logic with ontology which makes it difficult to accept even his carefully worded claim that his *Logic of Preference* is "semi-extensional" with the logic of propositions. Surely, to claim that the absence of a state of affairs is suggested by a negated preference is to extend logic of propositions beyond what it is designed to deal with. In this respect, the difficulty is compounded by the fact that von Wright interprets his meta-theorems (1) and (2) above with the introduction of the modal conditional "may," which again is not the idiom which the logic of propositions deals. Consequently, even the so-called semi-extensional status of the logic of preference with the logic of propositions is subject to contention.

Von Wright's next important distinction is that between "indifference" *between* states, as in saying pIq, and "indifference" in itself with regard to some individual state, say as in pIp. In the case of the former, he defines indifference as not preferring the state: p and q, to its negated contrary, e.g., '-(pPq)&-(qPp)'. In the latter case, indifference is defined in terms of the single state p, so that: '-(pP-p)&-(-pPp)'.[40]

However, as defined above, indifference does not express an "unconditional" indifference as one must have in a logic which concerns itself with unconditional preference. For this reason, he endeavors to redefine the notion of indifference in a "stronger" sense, introducing thereby the notion of "value-equality." The problem here stems very simply from the realization that indifference between two states p and q, that is the relation expressed as -(pPq)&-(qPp), can be considered as holding only in *some circumstances*, if it is assumed that there are three states composing the "all circumstances" requirement of the logic, i.e., p, q and r. In such a situation, all that can be safely said is that under *some* circumstances, i.e., apart from r, it is not the case that p is preferred to q. The "stronger" sense of indifference, however — what von Wright above calls (value-) equality — is defined as where "...the state p is value equal to the state q..." and means "...under no circumstances is the state p&-q preferred to the state -p&q, or *vice versa*." States which are value-equal are necessarily indifferent, but it does not hold that indifferent states are also value-equal.[41]

Value-equality, symbolized by 'E', is shown by von Wright to be clearly symmetric and transitive.

---

[40] Von Wright, Georg Henrik, *The Logic of Preference*, p. 52.
[41] ibid., pp. 56-57.

Of the various value-equal relations he discusses, one of the most interesting involves the introduction of the r variable, which is explained as a state said to be different from p and q. He argues very succinctly that if p&r is value-equal to q&r, and p&-r is value-equal to q&-r, then p must be value-equal to q.[42]

Von Wright continues toward the close of his investigation to discuss the way in which E-tautologies can be demonstrated, as well as their consistency proven through an adaptation of his diagrammatic method. He closes his treatise with a fascinating presentation of a new logic, where P- and E-expressions are combined to constitute a "Logic of Preference and Value-Equality."[43]

This last phase of von Wright's work deserves careful scrutiny since it manifests the accumulative outcome of his entire investigation. As such it exemplifies the most advanced degree of complexity his logic attains, at least within the confines of *The Logic of Preference*.

First, a number of restrictions are placed upon the distribution of truth values for P- and E-constituents in PE-expressions.[44]

(i) The constituents of form w_1Pw_2 and w_2Pw_1 cannot both be assigned the same truth-value of truth.

(ii) Given the n-1 constituents of a sequence of constituents w_1Pw_2, w_3Pw_4, $w_{n-1}Pw_n$ are assigned the value 'true,' then n-1 must have this value as well.

(iii) Where w_1Ew_2 is assigned the value 'true,' then any other PE- expressions either containing or derived from w_1 and w_2 by substitution, must have the value (true). This would satisfy the requirement of symmetry and transitivity respectfully for E.

An interesting proof of a PE-tautology by means of the diagrammatic method is offered by von Wright to illustrate the total integration of the logics of preference and of value-equality. Thus, given the expression: (pEq)&(pPr) → (qPr), where each dot in the diagram below represents a possible constituent distribution relative to the formula given:[45]

---

[42] ibid., pp. 58-59
[43] ibid., p. 61.
[44] ibid., p. 62.
[45] ibid., p. 65.

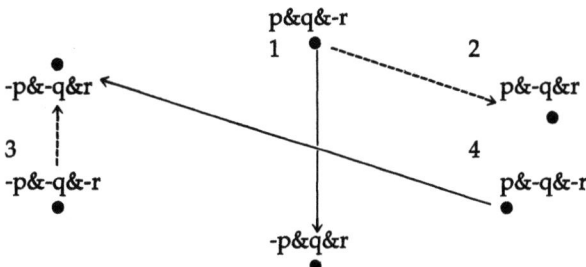

Arrows 1 and 4 both illustrate that state p is preferred to state r. Since -p&q&r is value equal to p&-q&r *given that* pPr, then dotted line 2 is required going from p&q&-r to p&-q&r. For the same reason, dotted line 3 is required from -p&q&-r to -p&-q&r. Both dotted arrows represent how q is preferred to r, and thus illustrate in a decisive manner the tautologous character of the above expression.[46]

Von Wright's return to the task of developing a logic of preference almost a decade later exhibits an even greater degree of sophistication as to all that which preference can possibly mean, while retaining his basic conception of intrinsic preference. For example, in "The Logic of Preference Reconsidered," 'xPy' is said to mean "possible (total state of the) world x is preferred (more liked, considered better, ranked higher) than possible world y." More interestingly, he demonstrates a greater responsibility for justifying transitivity in regard to preference-relations than in his views on the transitivity of preference in *The Logic of Preference*. In this regard, he suggests a most intriguing solution to the question of how one is to justify saying that where s is preferred to t, and t is preferred to u, therefore s is preferred over u. He observes that transitivity holds if the "coherence" of his system of preference is to remain intact. In other words, the transitivity of preference-relations holds only if one is dealing with a system of holistic preferences, within the same "horizon" of states of affairs. He goes on to suggest that under the circumstances it may be best to require that coherence become a *criterion* of rationality, so that the transitivity of preference in his sense may be assured. He characterizes this move as the "logifying" of nature, which makes possible the productive investigation of preference

---

[46] ibid., p. 66.

in the intrinsic sense he finds so important. Coherence becomes a means of asserting the *unity of time* for a universe of discourse involving preferences in his sense. For if it were not the case that *s* is preferred to *u* in the above example, then a different world-state has been introduced, and, consequently, a new horizon of states of affairs must be considered.[47]

Von Wright's remarks on the role of coherence in safeguarding the transitivity of the relation of preference find an interesting parallelism in some of Rescher's observations in *The Coherence Theory of Truth*. Though Rescher aims his discussion on this point at the issue of deciding what objective conditions would come into play for determining which "maximal consistent set" (m.c.s.) is the preferable, there are still some serious ramifications to consider in what he says relative to von Wright's position on the transitivity of the preference-relation. Specifically, Rescher argues that one would prefer that m.c.s. whose propositions exhibit the greatest degree of truth, where truth is taken to be a function of consistent interaction by the propositions within any given subset. It is seen here that Rescher conceives of the preferable as a judgment which results from assessing the "point of truth" of propositions within a mutually consistent, i.e., *coherent*, subset.[48] However, in applying what Rescher says here, it is evident that preference transitivity relies upon or results from the *logically consistent integration* of the expressed preference with *other* expressions within a systematic ordering, and that it is not dependent upon or due to the *intrinsity* of the preference itself, though the latter is basic to von Wright's view. Ironically, von Wright's proposed solution in his later essay would force him to adopt a criterion, namely coherence, which by its very nature entails the introduction of extrinsic determinations, i.e., how the various propositions within a unity of discourse logically relate to each other, so as to preserve the cohesiveness of his logic of preference. Surely, without the transitivity relation, his logic would be woefully limited, if not impossible. Yet, the solution he proposes would minimize in a very important respect the importance of considering preferences solely in terms of their intrinsic nature.

Published criticism of von Wright's earlier work is not voluminous. However, in their celebrated article, "On the Logic of "Intrinsically Better,"" Chisholm and Sosa assume a position which illustrates a highly conservative rendition of the notion of "intrinsic goodness," in comparison to what can be

---

[47] Von Wright, Georg Henrik, "The Logic of Preference Reconsidered," p. 150.
[48] Rescher, Nicholas, *The Coherence Theory of Truth*, Oxford at the Clarendon Press, 1975, p. 84.

seen in von Wright's work, for example. This is, that intrinsic goodness is taken to mean "that which yields pleasure." What does not yield pleasure is intrinsically bad, and nothing further can be inferred. On the surface this may seem close to the kind of presupposition one would find in von Wright. However, there is a profound cleavage at the very outset between the sort of thing von Wright is advocating, and what one finds in this work. Chisholm and Sosa reject the hypothesis that an intrinsically good state of affairs, for example, *in itself implies* that its negation is (intrinsically) bad. Rather, they maintain that a state of affairs, whether exemplified or not, which is said to be intrinsically good, implies *only* that it yields pleasure. Hence, it is plausible to characterize the position as "conservative," in the context of von Wright's position. The latter, by contrast, allows for the forementioned inferences. Moreover, Chisholm and Sosa stipulate that the only thing that is intrinsically good is pleasure. In this connection, they also say that any state of affairs which involves pleasure is better than a state of affairs that involves (or results in) displeasure, but that there is no *implication* between these two kinds of states of affairs.

Within the framework of these hedonistic assumptions, Chisholm and Sosa proceed to reject several fundamental principles dealing with the logic of preference, as found prominently in von Wright's work.

First among these is the primary principle which holds that p is intrinsically good if and only if it is preferable to its negation, i.e., $p \equiv pP\text{-}p$. The authors point out that supposing p designates the case where there are no unhappy egrets, then the negation of p, which is say the state of there being unhappy egrets is not better that there being no unhappy egrets. For the absence of displeasure does not in itself rate the universe a plus, relative to the hedonic assumptions already accepted. In other words, Chisholm is arguing here that nothing can really be inferred from the state of the absence of displeasure. Thus, the above principle is clearly false, at least when applied to this case.

The second preference-principle to be rejected is the one which says that p is bad if and only if the negation of p is preferable to p, or $Bp \equiv \text{-}pPp$. Here again one would say that the state of their being no happy egrets, p, is not intrinsically bad, so as to justify saying that the state of their not being no happy egrets is better that it.

Similarly, the third principle to be rejected is that which claims: $pPq \equiv \text{-}qP\text{-}p$. Again, Chisholm and Sosa use the example where though it be the case that there are happy egrets, p, is more preferable to there being stones, q. For the state of their being no stones is neutral, hedonically, and thus not better than there being no happy egrets.

The fourth principle rejected is the one which claims that the preference of p over q is equivalent to the conjunction of p and -q preferred over the conjunction of q and -p, or more precisely: pPq ≡ (p&-q)P(q&-p). One way of disposing of this principle is to argue that where p is the state of there being stones, and q is the state of their being no happy egrets, then whereas there being stones and happy egrets is preferable to there being no happy egrets and no stones, this in no way means or implies that it is better that there be stones to there being no happy egrets. Again, the basic point here is that there being no stones is hedonically neutral, so to speak, from which the equivalence relation as expressed above need not follow.

The last principle to be falsified involves the subtle distinction between "indifference" and "neutrality" between states. The principle in question says: "every state of affairs is either intrinsically good, bad, or indifferent." Here indifference is usually explained as -(pP-p) & -(-pPp). Chisholm and Sosa see the falseness of this principle in the fact that it is neither intrinsically good nor bad that there are no unhappy egrets, but it is still a state of affairs which is intrinsically better than its negation, hence it is not an indifferent state of affairs. These authors see an advantage in using the kindred notion of neutrality instead of indifference, so as to say that a state of affairs is neutral, if it is neither intrinsically good nor bad, i.e., Np ≡ -Gp & -Bp.[49]

Clearly, these authors propose an approach to a logic of preference which puts into question the very notion of transitivity assumed by von Wright at the outset of his work. As noted earlier, von Wright proceeds on the assumption that a pleasing state of affairs, and therefore one which is intrinsically good, is preferable to one which is not pleasing, and consequently intrinsically bad. Without the consequential relation between the intrinsically good and bad, which is in essence von Wright's relation of preference, no relation of transitivity is possible.

Von Wright's sensitivity to the need to justify formally the transitivity of preference-relations constitutes only another point of departure from his work in *The Logic of Preference*. This is to say that he comes to realize through the for-

---

[49] Chisholm, Roderick M. and Sosa, Ernest, "On the Logic of "Intrinsically Better,"" *American Philosophical Quarterly*, Vol. 3., No. 3, July 1966, pp. 244-246.

mal analysis he performs in "The Logic of Preference Reconsidered" that preference — in the holistic sense he favors throughout his work — though asymmetric and transitive, does not provide a means of generating a linear preference ordering. This result has profound importance. It illustrates that the assumed relation between preferences as involving absolute goodness and the relative notions of the better and same, as encountered in the work of Soren Hallden for example, is not as close as von Wright had supposed in his earlier work. This is the interesting insight which emerges from his later essay, and it may well be one reason why he calls for a concentrated effort by future investigators to disentangle the confusion which seems to beset the notion of preference on an intuitive level of analysis. For it does not seem to him readily possible to proceed from relative judgments, passing from saying that since x is better than y, and y is better than z, to the claim that: x *must* be good, and z *must* be bad, in some pervasive and intuitively certain sense of these terms.

However, to see how von Wright arrives at the above conclusion requires that one follow the unfolding of his later argumentation, with careful attention to detail.

Von Wright begins by concentrating upon holistic preferences, *under given circumstances* C_i. The circumstance C_i he explains in two distinct definitions. First, D1. says that "... s is preferred to t under the circumstance C_i, *if and only if, every C_i-world which is also an s world but not a t-world is preferred to every C_i, if and only if, every C_i-world which is also a t-world but not an s world. ...*" Thus in the case in point, von Wright asks us to consider the selection of s over t to be such that there is *no* part of preferred state s which is part of t, as D1. above stipulates. He pictures the two states s and t as represented by two dots on a given line, so that s is to the left of t. And thus, any state which is depicted as being to the left of another is a representation of a preferred state. Another way of expressing preference *under given circumstances* is through his second definition of C_i. In D2. one has "... s is preferred to t under the circumstance C_i, *if and only if, some C_i-world which is also an s-world is preferred to some c_i-world which is also a t-world, and no C_i-world which is a t-world is preferred to some C_i-world which is an s-world. ...*" With whichever of these definitions one adopts, the P_i-relation, i.e., the preference under circumstance C_i-relation, is found by von Wright to be asymmetric and transitive. In the pictorial model he presents, this asymmetry and transitivity are clearly seen, given the left/right convention mentioned above.[50]

---

[50] Von Wright, Georg Henrik, "The Logic of Preference Reconsidered," pp. 154-155.

Yet this same model shows that there are no grounds for holding value-equivalence between preferences. For example, where one considers the line:

```
     a    b  c  d    ·   ·   ·
           ·  ·  ·
    ─────────────────────────
```

where s represents the disjunction of all the total states a to d, t is the state represented by c, and u is represented by d, then whether one applies definition D1. or D2. above, no preference-relation can hold such that $sP_c t$ or $tP_c s$, or $sP_c u$ or $uP_c s$. For by means of the Principle of (Value-) comparability which states $(xPy) \equiv [(xPz) \lor (zPx)]$ — assuming the relation of transitivity and a third state designated as z — it can readily be seen that diagrammatically the left/right convention allows for *no* linear ordering of preferences among s, t and u.[51]

Von Wright finds that the same result occurs where one takes into account the other type of holistic preference, i.e., preference in the intrinsic sense, considered under the *ceteris paribus* condition. Von Wright's intricate proof on this score shows how the conjunctive totality of preference-relations here, though symmetrical and transitive as in the above case, do not constitute a linear ranking order within the same or given state-space. This suggests that von Wright is sensitive to a deep-rooted limitation in his conceptual analysis of preference, which renders his work incomplete. Surely, the ranking of preferences constitutes, or at least should constitute, a major aspect of any such analysis.[52] Yet von Wright seems to be realizing that in the circumscribed total world states on which he focuses his attention there is no possibility of illustrating the hierarchic ordering among preferences. Interestingly, it will be seen that some of the major efforts in this area which follow after von Wright's seminal work, e.g., N.Rescher and R.Jeffery, specifically concentrate upon formalizing the *ranking* of preferential states of affairs. For von Wright, however, the limitation he encounters on the point of linear ordering of preference is unexpected in the sense that intuitively he expects his logic of preference to enable one to express levels of preferring, which in turn would be indicative of levels of better or worse, and consequently of good and bad.

The above also brings to von Wright the realization that one should be aware of two notions of indifference, one stronger than the other. The first, which is the weaker, is simply reflexive and symmetrical, but not "generally"

---

[51] ibid., p. 155.
[52] ibid., p. 155.

transitive. This is expressive of preference, whether P is considered under the *ceteris paribus* condition or the given circumstances condition, such that: -(sPt) & -(tPs). Here it is important to point out, as von Wright himself cautions, that it does not necessarily follow that because s is not preferred to t, and t is therefore not preferred to s. Herein lies the need for limiting the transitivity relation in this version of preferential indifference. The second (stronger) definition of indifference, which is transitive and reflective of value-equality, states that s is *ceteris paribus* strongly indifferent to t, in state space S, if and only if no s-world is "*ceteris paribus* under the circumstances C_i, *preferred to any t*-world, nor any t world to any s-world." In this, the most exclusive version of his second definition of indifference, von Wright is most careful in making sure that the transitivity relation is not restrained in its reciprocal function.[53]

Surely, von Wright's logic is an achievement in philosophical logic. Apart from the objection noted on the conceptual presuppositions of what he says, it cannot be denied that he is the very first to see the logic of preference as a uniquely philosophical concern. Indeed, his early decision to pursue a logic of intrinsic preference, as opposed to a logic of preference which involves notions of economic utility and probability, is a pioneering step in the direction of seeing such a logic as a "philosophical" inquiry. In the context of what came before, von Wright is not faced with the general problem which confronted Ramsey, Davidson and Bohnert, namely the difficulty of fitting a proposed theory of preference, usually articulated on the basis of a probability theory, to the facts of the physical world. As has been seen, von Wright seeks to explicate logically an intuited conception of preference, independently of any extensional considerations. However, the issue which does plague his attempt is more internal in nature, in the sense that he sees that his own logic of preference falls short of providing a means for determining levels or degrees of preference, which in turn would reflect degrees of the better and worse. In a sense it can be said that the logic of preference which is ultimately evolved does not reflect the delicate levels of discrimination which he feels a general theory of value should reflect. However, unlike Hallden, and most of his predecessors, von Wright works with what he takes to be the "sense" of natural discourse. He attempts to render in terms of the terminology of propositional logic the relations which he intuits as contained within discourse involving preferences.

---

[53] ibid., p. 156.

Chapter 8.

Hochberg on the Logic of "Extrinsic Epistemic Preferability"*

As it has been seen, efforts at evolving viable calculie of preference have been pursued mainly from two distinct directions. One seeks to formalize within a modal logic the intension of preferential expressions, considered in terms of object having intrinsic value. The other proceeds by taking preference in an extrinsic sense, and then axiomatizing preference-principles. However, beyond a few elementary relationships of preferring, e.g., (pPq) → -(qPp), (pPq & qPr) → pPr, etc., there is virtually no general consensus within any one of these two traditions of analysis as to what counts as a "genuine" preference-principle.

In the context of this setting, one has Professor Hochberg's contribution, which is unique in its way of attempting to combine both of the above ways of developing logics of preference. By expanding certain formal definitions of a general value calculus for preferences, considered intrinsically, he also seeks to accommodate "the essence of extrinsic preferability. ..."[1]

Apart from the virtues and shortfalls of his suggestions, his efforts come to magnify the subtle difficulties involved in attempting to have a single calculus which is descriptive of both an individual's personal likes and dislikes-- having no objective determinations (i.e., intrinsic preference), and that same individual's experiential judgment regarding the relation of means to ends (i.e., extrinsic preference) in the physical realm.[2] These difficulties reflect the central issue of the adequacy of any logic in explicating the formal implications of sentences expressing subjective as well as objective state of affairs. Conse-

---

* Excerpted from remarks presented at the International Meeting of the Symbolic Logic Association, Patras, Greece, Summer-1980.

---

[1] Hochberg, Gary M., "Extrinsic Epistemic Preferability," *Philosophical Studies*, Vol. 23, No. 1-2, Feb. 1972, p. 83.

[2] ibid., pp. 76-77.

quently, one vitally interesting aspect of Professor Hochberg's work is the way he considers the introduction of factors concerning epistemic relevance as a means of deepening the explanatory power of an intentional logic of preference, which would then accommodate preferences extrinsically.

I

The definitions Professor Hochberg sets forth at the outset of his paper are taken as the ground from which a logic of extrinsic preferability "may" be articulated. It is important to stress the sense of "may" here, since Hochberg is sensitive to the tentative character of his remarks. Basically, his strategy is to present a set of four definitions regarding intrinsic preferability in a general value calculus, i.e., definitions covering preferential *indifference, sameness, neutrality,* and *goodness,* and then to offer a series of definitions governing what he calls "extrinsic terms", e.g., the extrinsically *reasonable, acceptable, favorable, indifferent, gratuitous,* and *unacceptable.* From this, he attempts to set forth through the introduction of epistemic considerations a logic of the extrinsically preferable, which contains for its inner core the logic of the intrinsically preferable, outlined in the first group of definitions he presents above.

Briefly, Hochberg's definitions in the general value calculus run as follows:[3]

(1) 'p is indifferent': (Ip) - It is not the case that p is preferred to -p, and it is not the case that -p is preferred to p: (-(pP-p) & -(-pPp)).

(2) 'p is the same in value as q': (pSq) - It is not the case that p is preferred to q, and it is not the case that q is preferred to p: (-(pPq) & -(qPp)).

(3) 'p is neutral in value': (Np) - Relative to some particular q, which is indifferent, p has the same value as q.: ((q)(Iq & pSq)).

(4) 'p is good': (Gp) - Relative to some particular q, which is indifferent, p is preferable to q.': ((q) (Iq & pPq)).

Apart from the difficulties of accepting the intelligibility of (3) and (4), in that "neutrality" of value does not seem to be sharply distinct from "indifference", and p's "goodness" does not seem clearly different from the simply "preferable" — which he does not define in any sense, one can provisionally entertain

---

[3] ibid., p. 76.

definitions (1) through (4) for the sake of unraveling Hochberg's calculus of extrinsic preference.

Yet at this early juncture one sees how intuitive notions involving intrinsic preferability, are in essence nebulous. For example, it would seem that neutrality has to be somewhere beyond the indifferent, preferable, and non-preferable. In fact, it can be argued that the neutral should have some relation with the irrelevant. Moreover, it does not seem to be very clear how the good is the preferable in relation to the indifferent. With (4), Hochberg is slipping very close to some kind of means-end concept, which he explicitly reserves for his discussion of extrinsic preferability. Surely, to say that the intrinsically good is the preferable over the indifferent carries some suggestion of effects being considered in the background of the analysis. From a broader perspective, it is important to note that within the evolution of logics of preference, those who have attempted the analysis of preference from an intrinsic viewpoint have faced the same seemingly paradoxical situation, which is having to deal with the term's intensions without reference to any fixed context, while also contending with the fact the extrinsically preference *must* involve reference to agent, action, and time. Significantly, if preference in an intrinsic sense is to be construed in terms of personal desires and wants, then how is it to be distinguished from mental dispositions? Yet, preference in an extrinsic sense appears to involve reference to instrumental means directed at a particular end. Thus, the aforementioned paradox centers around a tension between the subjectivist intrinsic sense of preference, and preference in an objectivist extrinsic sense. One seems to have here the classic subject/object antithesis, which makes Hochberg's approach all the more interesting.

His definitions of extrinsic epistemic terms need not be presented in their entirety. A few representative definitions will suffice:[4]

(1) 'P is reasonable' (Rp) - Believing p is preferable to withholding p. (BpPWp).

....

(3) 'p has some presumption in its favor.' - Withholding p is preferable to believing p. (WpPBp)

(4) 'p is epistemically indifferent'. (Ip) - Withholding p is preferable to believing not-p. (WpPB-p)

---

[4] ibid., p. 77.

These definitions exhibit an extrinsicity in the way they are meant to involve the role of publicly confirmable evidence is the determination of what counts as reasonable, favorably presumed, and indifferent. Hence, these are the only definitions given in terms of the epistemic operators of believing and withholding. Presumably, Hochberg is working with some notion of belief in an extensional sense, possibly modeled according to Peircean lines. The allusion to the preferable in these definitions is understood in terms of the intrinsically preferable brought out previously. Thus, it may be said that the extrinsic epistemic terms he sets forth already have some relationship to intrinsic preferability, in that the preferential seems to be constituted by that which is wanted or desired after the extrinsic epistemic conditions have been met.

However, as Hochberg correctly observes, extrinsic epistemic preferability requires reference of means to ends, as well as reference to an agent and a time of action. These aspects of the analysis do not play a role in considerations of intrinsic preferability, which underscores the difficulty in accepting his first set of definitions governing such preference.[5]

To fail to bring into play the role of means to ends, and agent to time, could commit one to a logic which, according to Hochberg, is inadequate for explaining the facts which are involved in an extrinsic preference situation. Here he makes an astute observation in that one cannot evolve an extrinsic logic of preferability by simply using the conjunction of states of affairs, or by considering their mere sum. This is to say that to define extrinsic preferability in terms of saying that p is extrinsically preferable to q relative to r, or '(p exP q)Rel r', does not capture the richness of preference extrinsically. For example, if the relative situation is "getting to Boston", and 'p' means 'having a car', and 'q' means, 'having one million dollars', then one cannot simply capture the subtlety of extrinsic preferability by saying that, therefore, pPq, relative to r. For the *real world* context may very well support the view that q is preferable to p relative to r, in this case. Which is to say that having the million dollars may be more conducive to getting to Boston than having the car. Thus, one cannot operate solely on the premise that the value of whatever state of affairs is preferred directly results from the *mere sequence* of the states of affairs the preference involves. *More* needs to be said, according to Hochberg, with respect to the *relevance* of the state of affairs to the condition of preferability.

In a sense, Hochberg is pointing to the narrowness of the primitive operator 'and': '.' in its ability to capture the linguistic nuance which is influ-

---

[5] ibid., p. 79.

enced by the epistemic conditions the person doing the preferring finds himself in. As a result of the above insight, Hochberg turns next to the examination of epistemic relevance, so as to return to the logic of preference with a much more powerful conception of the factors relevant to preference in an extrinsic sense. According to Hochberg, relevance is the key notion which gives a needed sense of epistemic depth to preference, considered extrinsically. However, "epistemic relevance" is very difficult to pin down, from Hochberg's viewpoint, since it defies analysis in terms of "mere conjunction", as noted above. Yet, it is the only connecting point by which one can go from an intrinsic to an extrinsic logic of preference. In this connection, he proceeds by observing that the notion of epistemic relevance plays a key role *only if* some means can be found by which it can be argued that there is "greater epistemic status" attached to a proposition, relative to a particular preferential context. This is to say that Hochberg is searching for a way of expressing the idea that in the case of extrinsic epistemic preference, what is preferred is so preferred because, relative to some particular state of affairs which the preferential conditions require be brought about, that state of affairs acquires a greater epistemic status for the agent through some action he performs, which comes down to saying that the relevant state of affairs becomes more "certain" or reasonable as far as its imminent realization is concerned.

To secure the requisite idea of epistemic relevance, Hochberg alights upon the complement of the notion of "bringing something about." He contends that it is the notion of "to fail to bring about a state of affairs" which works to trigger a logic of extrinsic preferability.

Turning first to some new definition of extrinsic epistemic value, Hochberg offers the following as the new underpinning for a truly extrinsic logic of preference:[6]

> (1) 'p is extrinsically indifferent relative to r' ((exIp)Rel r) - It is not the case that failure to bring about p in conjunction with the attempt to realize r is preferable to failing to bring about non-p in conjunction with the attempt to realize r, and it is not the case that failure to bring about non-p in conjunction with the attempt to bring about r is preferable to failing to bring about p in conjunction with the attempt to realize r. (-(-Cp&r)P(-C-p & r) & - (-C-p &r) P (-CP & r))

---

[6] ibid., p. 81.

(2) 'p is extrinsically the same as q, relative to r' ((p exS q)Rel r)) - It is not the case that failing to bring about p in conjunction with the attempt to realize r is preferable to failing to bring about q in conjunction with the attempt to realize r, and it is not the case that failing to bring about q in conjunction with the attempt to realize r is preferable to failing to bring about p in conjunction with the attempt to realize r. (-(-Cp & r) P (-C-p & r) & -(-C-p & r) P (-Cp & r)).

....

(4) 'p is extrinsically preferable to q, relative to r' ((p exP q) Rel r)) - Failing to bring about q in conjunction with the attempt to realize r is preferable to failing to bring about p in conjunction with the attempt to realize r. ((-Cq & r) P (-Cp & r)).

(In all of the above definitions, 'C' is to be read as the operator: "to fail to bring about.")

Hochberg contends that there is an analogy between the notions of "bringing about" and "failing to bring about", and the notions of "believing" and "withholding," discussed in connection with extrinsic epistemic terms. Thus, very simply, failing to bring about one state of affairs is tantamount to the withholding of a proposition, say p. Similarly, bringing about a state of affairs is analogous to believing proposition p. Thus, Hochberg sees a connection between operators which work on propositions ("believing" and "withholding").

To encapsulate Hochberg's argument one can simply reflect upon the important points he has made thus far. First, his introduction of the "intuitively" valid general value calculus is intended to provide a version of relations of preference in an intrinsic sense. Second, his introduction of extrinsic epistemic terms is designed to introduce considerations requiring "public confirmation". This leads to the third aspect of the analysis, which is the presentation of extrinsic epistemic values which involve actions which confer a greater epistemic status on the agent's position in securing what is preferred. However, the third aspect of the analysis ends with considerations involving *praxis* which is represented by the operator: "failing to bring about." Consequently, Hochberg introduces a new operator, i.e., "withholding", which is a way of handling propositions expressing preferences, and is the counterpart of "failing to bring about."

Something further needs to be said concerning Hochberg's understanding of the parallelism between "failing to bring about" a state of affairs, and the "withholding" of acceptance for a proposition. The move he has made is crucial, and further elaboration is important for the purpose of clarifying his entire position. Though he himself does not supply the details, it appears that the notion of failing to bring something about is an *action* one undertakes as a result of deliberate understanding of what the effects will be if something is not done within a specific physicalistic context. On the other hand, "withholding" is an *operator* which works on propositions, in the sense that one does not accept the truth of sentence *a*, say because certain undesirable results will accrue by doing so. Perhaps the notion of "withholding" can be understood best by considering its complement, i.e., "belief." Presumably, "belief" is taken in the Peircean sense of acting upon the meaning of a proposition, in that believing a means that one acts in such and such a way. Thus "withholding" requires that one not act upon a said sentence, for the purpose of bringing about some desired result. Outwardly, one seems to have the paradoxical situation that withholding acceptance on *a* is to "bring about" some effect, which would presumably be facilitating the acquisition of what is preferred. However, the paradox is dissolved by keeping distinct the two contexts of *praxis* and *epistemie*. This is to say that from a praxiological point of view there is an immediately recognizable difference between bringing about and failing to bring about. On the other hand, from an epistemic point of view, one can say that there is an immediately recognizable difference between believing *a* and withholding acceptance of *a*. Presumably, since epistemic insights precede praxiological events, it can be maintained that the notion of withholding acceptance of *a* underlies the notion of "failing to bring about." In this way Hochberg connects the epistemic element with the praxiological within a single analysis of preference.

Hochberg insists that the clarity of this parallelism will become manifest as soon as he makes his final move, which is to set forth his last four definitions of extrinsic epistemic preferability. For brevity he concentrates on explaining only one of these four definitions. Consequently, it is this which will constitute the substance of the final point of the present discussion. Hence, he defines the extrinsically preferable as:[7]

(4) 'p is extrinsically preferable to q, relative to r' ((p exP q) Rel r)) - 'Withholding q relative to r is preferable to withholding p relative to r' ((Wq & r) P (Wp & r))."

Hochberg endeavors to demonstrate the tenability of (4) by considering the case where: 'p' designates the case of reading and understanding the proof of a

---

[7] ibid., p. 82.

theorem of geometry, 'q' designates the case where one is told by an expert mathematician that the said theorem has a proof, and 'r' is the said theorem of geometry. Hence, using (4), one has the logical characterization of extrinsic epistemic preferability set out as withholding q: the expert geometrician telling the subject that there is a proof for the theorem, designated by 'r', is preferable to withholding p: the subject's own reading and understanding of the proof of the theorem.

Hochberg underscores the point that the epistemic status of r remains unchanged by the preference expression presented above. Rather it is the subject's own position which is enhanced toward the realization of r because of the preferability of the withholding of q. Hence, returning to the point which underlies his entire thesis, the object of a logic of extrinsic epistemic preferability is to secure a formal description for the advantage of one's epistemic position "as regards *realizing* the status of r," that is, the status of the relevant condition required for bringing about what is preferred.[8]

In summary it is to be noted that Hochberg's last definition contains the blending of a number of important elements. These are:

(1) The relevant state of affairs which sets the context for the preference, expressed by 'r'.

(2) Two presumably exclusive courses of action which are involved in the securing of r, these are propositions expressed by 'p' and 'q'.

(3) The "withholding" operator which serves to express the agent's disapproval of the expressed state of affairs.

(4) The withholding performed in (3) serves to *confer a greater epistemic status* on our understanding of 'r' whatever it may be taken as designating. Thus the epistemic component is shown to be involved with the extrinsic dimension of preference, and it would seem that Hochberg has convincingly secured his main point of presenting a systematized account of extrinsic epistemic preferability.

Moreover, it can be seen that he seems to have fashioned a transition from his definitions of intrinsic preferability to extrinsic preferability by simply introducing his notion of epistemic relevance. Consequently, on the surface it

---

[8] ibid., p. 82.

would appear that one could accommodate extrinsic preferability within a logic of intrinsic preferability, with appropriate modifications.

The main difficulty with Professor Hochberg's position centers around the viability of the notion of "conferring greater epistemic status upon" some recognizable end which a preference situation involves. What does it mean to say that one confers such a status upon a proposition by withholding — presumably one's assent upon — that proposition? The question really raises the knotty problem of causal determinacy. For by withholding one can never be absolutely sure what will not come about, within the natural order of things.

In this connection, it is worth noting that Hochberg's example of the geometrical theorem really does not serve to cover the complexity of empirical factors which are at work in the notion of means-ends preferability. For knowing how to work out the proof of a theorem which helps in improving the position of the agent relative to his ultimate mastery of the study of geometry, does not reflect the subtlety of how empirical factors manifest themselves in preference. Surely, knowing how to do a proof within the defined system of geometry may be a necessary prerequisite for mastering the subject matter, and as such the whole notion of "knowing how to go on" in geometry may hinge on knowing how to do the proof Hochberg mentions in his example. All this seems separate from the reality of physical events, with their apparent novelty and surprise. The point here is very simply that the example Hochberg uses is misleading in the sense that if the knowing of a proof is necessary for going on in learning geometry, then given that the object is knowing a proof geometry, it follows from the fact of the highly systematic organization of that discipline that one must master a proof on the way to completing the learning. The necessity here is analytic, in that it is due to the definitional requirements of geometry, and not because of the rigor of Hochberg's definition of extrinsic preferability. It can be noted in passing that his rendition of the element of epistemic relevance, designated by 'r', is incomplete if it is taken only in the sense of a theorem of geometry. This is to say that one does not first confer "greater epistemic status" upon a theorem of geometry, and thus enhance his own position, unless what is meant is that the knowing of the said theorem is conducive to some broader end, i.e., "knowing how to go on" in geometry. This is consistent with Hochberg's own insistence that it is the agent's own position which is enhanced by withholding one proposition rather than another.

The drawback of Hochberg's view can be illustrated by going back to the example he gave about going to Boston. If his definition of extrinsic epistemic preferability is sound, then it should be the case that with respect to the relevant epistemic condition of getting to Boston, r, an agent's withholding of the proposition regarding the having of one million dollars, rather than withholding the proposition of having a car, would insure the conveying of epistemic purport to the relevant element, expressed by r. However, his definition leads one almost to complete absurdity, since it is soon seen that the agent's position is not necessarily improved by withholding the former and not withholding the latter. The old problem, in essence, re-appears, namely in the real life situation of agent's, actions, and physical events, the bringing about or belief in having the one million dollars may confer a greater epistemic status on r, than believing that one should have a car. This only goes to show how preference in an extrinsic sense brings into play empirical factors which cannot be given a conclusive formal characterization in the manner Hochberg suggests.

Apart from its shortcomings, Hochberg's contribution to the field is vitally important since it focuses in a very direct way on how a dispositional attitude like want, and a presumed determinable extensionalism, seem to be sharply distinct. This is to say that in the criticism of Hochberg's thesis one finds that there is implicit in his perception of extrinsic preference certain physically stable reference points, i.e., the relevant end of the preferring, the two exclusive courses of action designated by 'p' and 'q', the physically determinable effect of withholding, etc. This can be summarized as the understanding that there is an objective physical realm "out there", elements of which can be designated extensionally, and which can be considered in terms of their causal efficacy when considerations of withholding come to play. However, this physical component, when allowed to manifest its novelty, reveals that apart from any epistemic understanding of the causal relationships it manifests, it cannot be pinned down in such a way that it can be formally characterized, so that if such and such is "withheld" the following *will* or *must* occur. Granted the results uncovered thus far, one sees how the withholding of assent or belief toward a proposition about the physical world does *not necessarily* confer to the agent a greater epistemic status, or (what amounts to the same thing) a greater certainty toward the realization of the end. Thus, it can be pressed that the "withholding" operator does not seem to work in the way Hochberg intends it to, and that, therefore, it does not follow that what one wants or desires in an extrinsic sense of preference is in fact necessarily explained or secured by introducing the notion of "withholding." Consequently, it can be argued that the evident contingency involved in the understanding of physical phenomena creates a difficulty in securing a link between the intrinsically desirable, however one may want to interpret the role of desire in the notion of preference, and the extensional parameters of preference, considered from the viewpoint of means to ends.

In view of the above, it cannot be claimed that Professor Hochberg has secured a comfortable transition from the idea of intrinsic preferability to the extrinsic preferability. The two senses of preference still seem mutually irreconcilable, possibly because they reflect the debilitating classical dichotomy between subjective-self and objective-world.

The significance of Hochberg's attempt lies in the implications which result from his failure to consider the richness of the notion of preference, as a "form of life" in the Wittgensteinian sense. This is to say that though he comes to the realization that extrinsically preference comes to involve the intervention of an agent in certain physical states of affairs, his effort is hamstrung partly because he does not consider the varying interpretations which that intervention can have. This is due to the fact that the effect of that intervening is influenced and controlled by the novelty of physical events. Moreover, what he aims to do is interpret within a logical structure the complex ontology of extrinsic preference. The problem, however, is that it is not clear whether the variables he uses to refer to the constitutive elements of that ontology range over things, events, or processes. For example, in definition (4) what does 'r' refer to? What is the denotation of "relevance" in this definition? Also, what kind of event is a "withholding"?

In essence, Hochberg has not come to grips with the question of whether when dealing with a preference one is concerned with a physical object or mental event, in whatever sense one wants to identify these. By failing to set down a clear statement concerning the kind of entity he is dealing with, his ultimate definition of extrinsic epistemic preference does not reflect the uniquely social context which involves factors going beyond an agent's conscious intending. His general objective of evolving a logic of extrinsic epistemic preference from *within* a logic of intrinsic preference (involving the agent's desires and wants without reference to means-ends considerations), induces him to put the agent in the central role of the all-knowing interactor, conflicting with the environment. This gives to Hochberg's account of extrinsic preferability a requirement which demands that the agent consciously intends doing such-and-such so as to satisfy a desire and/or want. This picture is terribly restrictive, however. One may not know beforehand what he should prefer. Furthermore, in an extrinsic sense, the agent may come to "discover" what he should choose or not choose, on the way toward securing the end which he needs to realize. The point here is simply that Hochberg does not seem to have correctly understood the role of the agent in extrinsic preferability, and this tends to threaten the supposed centrality of "epistemic relevance" in extrinsic preference as conceived by him.

By way of positive suggestion, it may well be that one can secure a logic of preference by entirely rethinking the role of the agent in preferring. This is to say that it may be more helpful to get away from the subject/object split completely, and think of the purely instrumental factors which the concept of preference involves. Hence, rather than proceed from a general value calculus, as Hochberg does, it may be more productive to simply give a descriptive account of the agent's choice selection in a testing situation. Though this approach, previously suggested by R.M. Martin, forgoes any concern regarding intrinsic preferability, its virtue lies in its capacity for discovering patterns of selection which an agent adopts at particular times, within specified contexts.[9] In this way one has a series of pragmaticized relations between the language user and the extension of the expression used to express his choice. This particular methodology does not involve the complexities which are incurred through the use of the condition of epistemic relevance, as encountered in Hochberg's work. Rather than subscribing to a dichotomy which holds the subject/object division inviolable, the Martinian approach presupposes the notion of a communication event, which can be extensionalized in terms of its basic pragmatic components. Thus, preference acquires a broader conceptual basis, which takes into account the agent as sharing in a lived form of life with his environment, for the sake of effecting some end which is conceived as "needed" by him.

---

[9] Martin, Richard M., *Intension and Decision*, Prentice-Hall, (1963), Chapter II.

Postscript

There is as of the present no clearly universally accepted logic of preference, at least to the point of saying that it rest beyond serious fault as far as its applicability and/or internal consistency is concerned. The reason for this seems to be due partly to limitations in the very conception of the preferential, and how it is to be properly formalized. In this regard, von Wright's observation of the failure by philosophers to reach a consensus on what the logic of preference should be is correct, since there is discord by analysts as to how one should properly conceive of the preference relation in and of itself.

However, there is an additional factor to the above equation, which is that this disarray seems also to be linked to one's choice of the method for philosophical analysis. Typically, those adhering to a more scientific, and thereby to a more extensionally orientated inquiry, evolve logics of preference which prove to be restricted in their application to a broad range of phenomena. This was seen to be so with Ramsey, Davidson and Martin. In these cases there is an acute awareness of how the logical model falls short of conveying the richness which preferences seem to convey in the context of ordinary discourse. Thus, the extensional approach seems keenly sensitive to the limitation which any logic of preference has with respect to the variety lived experience presents. On the other hand, those engaged in a more intuitive approach to the analysis of preferences seem caught up with an end result which is of such a diffuse and diaphanous character that one wonders what their logics are meant to be about. This is to say, the methods of analysis which rely heavily upon the intuitive understanding of the intrinsically good in itself being essential to a conceptualization of underlying preferences, seem to evolve logics of preference which are unclear precisely as to what their logics are suppose to formalize. Here questions of ontology and precise reference are crucial, though unanswered.

More recently, John D. Mullen, taking a line of criticism which unilaterally opposes the viability of every attempted formalization discussed in this study, has challenged the very feasibility of a logic of *prohairetics* in "Does the Logic of Preference Rest on a Mistake?"[1] Mullen summary dismisses all such works by citing the evident disarray in the ranks of analysts when it comes to

---

[1] Mullen, John D., "Does the Logic of Preference Rest on a Mistake?", *Metaphilosophy*, 10, (1979).

agreeing upon even the most elementary features of the notion of preference. What else can such incongruity mean than that philosophers really do not have a legitimate subject matter to concentrate upon when it comes to the formal analysis of preference-expression? Quoting *verbatim* von Wright's statement of dismay over the clear failure by scholars to agree upon even the most rudimentary principles for a logic of preference, Mullen proceeds to identify the two basic flaws of all efforts to evolve logics in this area. The latter are conceived by him as being the reasons why *all* these attempts are fruitless.

Mullen's two-forked attacked upon *prohairetic* logics is designed to address procedural assumptions made in the course of evolving these logics. Thus, he begins by questioning the tenability of assuming that the principles of preference, so-called by analysts, are logically true, "... by virtue of the fundamental terms involved. ..." He finds it terribly irregular that the truth of such principles is a functions of such unsupportable assumptions. Secondly, Mullen goes on to cast doubt upon all the subjectivists efforts. For he does not see the rationale for holding that it is a sufficient criterion that such principles be judged acceptable because they are consistent with one's intuitive sense of reasonableness.[2] Either one or both of these assumptions seem operative in the overwhelming majority of cases where philosophers endeavor to produce logics of preference, and for Mullen this contributes to the demise of their efforts.

In presenting his argument, Mullen concentrates mostly on the logics presented by von Wright and Hansson, and proceeds to show how various theoretical contexts make different demands upon the conception of preferences in such logics. He insists that we cannot separate concept construction from theory construction, nor can concept evaluation be distinguished from a theory of choice. In line with this position he shows how, for example, the principle of the transitivity of preference is excluded from certain insights found in psychology. Moreover, he claims that *prohairetic* logics operate on the mistaken assumption that one can draw a sharp distinction between the essential and inessential properties of a concept as a matter of logic, whereas, in fact, modern theoretical physics has shown that this is not the case. Rather it is through *empirical* inquiries that one changes and alters what is deemed necessary for a theoretical concept. In this connection he notes that it is the decision of the theoretician which determines what is the essential property of a con-

---

[2] ibid., pp. 253-254.

cept, and this is not "a question of logic." Thus one should be willing to allow for those properties of a concept which ordinary logic would reject.[3]

The bottom line of Mullen's findings is that one cannot use intuitive reasonableness, nor logical consistency to evolve an essentialistic logic of preference, especially since it can be seen that such logics are inapplicable when it comes to specific scientific theories, either in the pure or in the applied sciences. The mistake of all logics of preference is that they are carried out in total isolation from theory construction.

The immediate value of Mullen's position is that it initiates new discussion into this important area of philosophical inquiry. Yet in reviewing his work one finds that he is prone to generalize about all logics of preference, so that one is lead to think that von Wright's logic is similar to Hansson's, and that both are virtually similar to every other attempt which formalizes preference relations. The failure of these efforts are thus seen as pervasive, given the shortcomings of the attempts mentioned above. Not only does Mullen remain insensitive to the different traditions of philosophical thinking which form the background of the various logics of preference, but the logical consequence of his argument is that philosophical inquiry as a specific discipline is itself suspect, since it will always fall short of explaining the conceptual idiosyncrasies of specific theories in various areas.

What Mullen's short essay also fails to appreciate is that in the literature in this field there is a constant effort to achieve a greater degree of flexibility, so as to capture the reality of choice in the real world. In many cases analysts are sensitive to the requirements in other areas, e.g., economics and decision theory, and attempt to achieve more powerful conceptions of preference so as to meet these demands. In the case of R. M. Martin's analysis of preference, for example, one has an attempt which works closely with the extensional requirements of experimental testing, so as to achieve an extensional pragmatics of preference. Even in the case of Georg H. von Wright's contribution, there is a clear indication of how logical analysis should *not* be absolutely separate from practical inquiries, since logic has an important clarifying function in its capacity to develop concepts.

One should also speak to the assumption implicit in Mullen's criticism that other disciplines, such as psychology and physics, can operate on their own in setting forth criteria for the preferable, and hardly need the often use-

---
[3] ibid., p. 255

less logics of preference developed by philosophers. The impression which his view gives on this score is strikingly dissolved by turning once more to Donald Davidson, and his seminal insights in *Decision Making, An Experimental Approach*. In his introductory remarks, Davidson observes that the purely introspective approach to the analysis of subjective utility is fraught with problems simply because one does not know how to translate the information it offers productively, so as to have numerical measures of utility and expectation.[4] It seems then that any psychological approach which employs introspective procedures must take into account the sort of analyses which consider the ostensibly observable factors involved in rational choice. Thus, an analysis of preference, and consequently the understanding of the concept of choice in decision theory, cannot be separated from the external observations involved in *actual* choice-making. The point then is that it is misleading to suggest that fruitful inquiry can be carried on here independently of experimental observation, since somehow psychological accounts of preference can be productive in and of themselves. For in Davidson's view, the assistance of the kind of insights provided by philosophers like F. P. Ramsey is needed, having introduced the idea that in matters of subjective utility one must consider factors of risk, thereby rounding out the *reality* of the actual choice-selection aspect of the agent's environment.

Thus Mullen's carelessness in generalizing about all logics of preference demonstrates a need for carefully understanding the crucial differences among the various logics of preference, whereby one soon uncovers a rich and varied texture of philosophical thinking. Ironically, this entire study is not at odds with the final conclusion Mullen draws, namely that no one logic of preference yet devised is totally beyond some fault. However, though Mullen's ultimate discovery is correct in spirit, the reasons he offers for the conclusion he draws are less than adequate. As it has become apparent in the investigations which were discussed, philosophers beginning with Ramsey show a progressively more intense interest for the way in which preference is "spoken of" in ordinary discourse. This direction of inquiry is unique, since it serves to base the philosophical study of preferences into something other than an insular intuitionism, contrary to what Mullen erroneously imagines.

Responding more directly to von Wright's observation, which is the springboard for Mullen's article, one sees how the issue goes deeper than just that of a difference in one's conception of preference, and therefore of a failure

---

[4] Davidson, Donald, *Decision Making, An Experimental Approach*, Stanford University Press, Stanford, California, 1957, pp. 10-11.

by thinkers to reach agreement on this ground. Fundamentally, there seems to be a problem here which emanates from the *way* in which philosophy is done, and this itself seems to be at the very basis of the disagreement as to how one is to properly formalize preferences. In saying this, one is not conceding to Mullen the thesis that *all* logics of preference rest on a mistake, meaning that all philosophical attempts endeavoring to derive them proceed without first explaining what the practical requirements for such a logic may be. It is not that logics of preference are based on this one common error, rather it seems more accurate to say that the variations are nurtured from different methodological approaches. It is this fact which contributes significantly to the above diversity of conceptualizations.

It is noted, however, that in his later essay, "The Logic of Preference Reconsidered," von Wright is found to conclude that because of differences in the very conception of preference, there is most likely not one single logic of preference, *per se*, but undoubtedly several. This point of view constitutes a change from the attitude which is found to prevail in his first work, *The Logic of Preference*, where von Wright appears certain that he has captured *the* logic of preference, considered intrinsically[5] Whichever view one takes here, it seems more charitable than the one offered by Mullen, though both are found to stand in opposition to the view given by R. M. Martin. The latter holds that there are actually no logics of preference to be had, but rather there are only formal analyses of specific situations involving choice. For example, one notices in Martin's analysis the emphasis upon the extensional aspects of a communication event involving preference. This is presented more in the way of showing how a relatively simple logic can be used to express certain relations within a context of choice. However, Martin is not intending to evolve a special logic of preference, which is fully independent of other notions he defines, such as 'belief' and 'acceptance.' Indeed, the thrust of Montague's criticism of Martin's analysis of preference rests on just this fact, namely that he (Martin) does not see the essential difference between one's believing that such and such is the case, and one's valuing such and such to be the case.[6] Rather, he interprets the notion of preference in extensional terms by alluding to belief and value, assuming that the latter can serve as a formal model for the former. Moreover, Martin's pronounced extensional approach refuses to accept the veracity of the

---

[5] Von Wright, Georg H., *The Logic of Preference: An Essay*, Edingburgh: Edingburgh University Press, (1963), pp. 17-18.

[6] Montague, Richard., "Richard M. Martin's Intension and Decision, A Philosophical Study," *Journal of Symbolic Logic*, Vol. 31, 1966, p. 101.

overview taken by von Wright, Hallden, and others. Rather, Martin allows only the possibility of formalizing individual or particular instances of choice, within a highly structured empirical context. At best, in Martin's view, one has only the formalization of expressions of preference, as sententialized expression taken in extension, but never a logic which eschews the relative factors of time, place, and action.

This division between those advocating the possibility of many possible logics of preference, and those claiming the contrary has a profound influence on von Wright's seminal insight that a logic of preference must lie at the core of a theory of valuation. For if this core, so-called, is illusory and not possessive of any critical mass, then theories of value are doomed to failure from the outset. Indeed, a vital portion of philosophical inquiry hangs in the balance, and it behooves one to achieve an understanding of precisely what von Wright means when he places the logic of preference at such a central position. Even were it to be shown that von Wright's view is correct, then there is still the problem of saying that since there are many possible logics of preference, there must be many different and yet equally plausible theories of valuation.

Consequently, what is meant by saying that the logic of preference is at the center of a theory of valuation? Apparently, for von Wright it means that it is the logic which along with a logic of permission, becomes the bedrock for a logic of action. This would in turn provide the proper grounds for restating and redefining terms and notions deemed necessary for utilitarian ethics. Though his perception of the importance of this field is evidently correct, in other writings he proceeds to say that one must be careful to distinguish between preference, as an expression of attitude, e.g., "liking x better than y," and the preferential as an expression of choice, e.g., "doing x as opposed to doing y." The latter being a very secondary sense of preference, if preference at all. More recently, he observes in his book, *Practical Reasoning*, that preference should be conceived as "the weighing of reasons," and that on this account preference constitutes *the reason* for overt action.[7] However, in this it is perhaps best that rather than referring to preference as an attitude of consciousness, one adopt the extensional view to the analysis of preference, with its reference to the ostensive elements in choice. This is more suited to the task of redefining concepts in utilitarian ethics, which is the ultimate practical application of the formalization of preferences. In summary, it appears that von Wright's approach to the notion of preference as one which involves attitudes, and his subsequent observation that the practical use of the analysis of preferences is to

---

[7] Von Wright, Georg H., *Practical Reason*, 1983, Cornell University Press, p. 57.

be found in its application to utilitarian ethics, seems to be a conflicting position to take. Surely, ostensive considerations in the activity of choice is basic to utilitarian ethical theory, for which reason it seems difficult to see how the more intuitionistic approach to the formalization of preference would be helpful.

Preference as the "weighing of reasons" implies a subjective rendition of the concept. It suggests a kind of judging quite apart from or prior to any overt activity, which is in itself observable in no spatiotemporal plane. Such a view is surely consistent with the kind of intuitionism von Wright accepts as the way of properly pursuing the logic of preference. This seems to convey the kind of phenomenological presupposition one finds in the work on preference by Chisholm, Sosa and Brentano. It is the precision of the concept which is at question here, for where or how is one to characterize the "weighing of reasons"? Is it descriptive, and, therefore, true or false of some sort of verifiable event? How is truth or falsity applicable in this context? These are the kinds of conceptual questions which an intuitionistic type of analysis cannot deal with.

The intentionalist viewpoint operates from the thesis that any choice is linked conceptually to some judgment which constitutes the reasons for the action the choice requires. Thus, any extensionalist analysis of preference, so-called, is, according to their view, presupposed by an intentional sense of preferring, which is necessarily primary and thus more fundamental. This is a view which von Wright is found to make much of in his latest writings on the way of analyzing preferences, asserting the importance of the intentionalist way of proceeding. However, can one surmise whether a thoroughgoing extensionalist approach, such as the one taken by R.M. Martin, for example, is intended to be descriptive of a state of judging, or as von Wright would say, of the "weighing of reasons?" Rather, it seems more accurate to interpret Martin's approach, as well as those of Ramsey, Davidson and Rescher, as being descriptive of the workings of the discourse about preferences. There is seemingly no reference to the dubious ontology of mental states in Martin's work, rather there is a direct reference to an extensional pragmatics, which endeavors to capture the essentials of the preference-relation within a context of objective testing and experimentation.

The pervasive problem with the intentionalist approach to the issue of analyzing preferences is the evident emphasis upon preference as a state of mind. Usually this is manifested by analysts saying that preference is truly synonymous with one's *"liking better"* that x be the case rather than y, - independently of any external factors influencing what one likes. Thus, preference from the intentionalist's perspective is expressive of or signifies one's dispositional attitude of *liking*. For Hallden, this is what the language of preference really means. The main reason why this approach becomes difficult to maintain is that it is unclear how one would hold true a preference claim which is founded on the idea of preference as a dispositional attitude. What sort of facts would confirm the preference claim in this sense as true. As subjective states, and though referred to by von Wright as "objective reasons," there is a certain inaccessibility to the verifying evidence except for the one doing the preferring. It surely cannot be that the logic of preference this approach defends is a logic solely about mental states. On the other hand, it cannot be a logic about ordinary discourse concerning preferences, since most writers in this tradition spurn any reliance upon ordinary discourse as a source which helps in the conceptualization of preference-relations.

Research in this area has also uncovered a further aspect of the notion of preference. This is that whereas one is tempted to accept von Wright's insight that a logic of preference is at the center of any theory of valuation, in an important sense preference itself presupposes valuation. This appears to be true especially in the case of the extensional approach, where preference is joined more with the idea of goal-seeking and relevant action. Surely here one's choice is guided by value judgments, and not conversely. Thus, ironically, von Wright's remarks concerning the centrality of the logic of preference are at variance with the kind of extensionalist approach which *alone* provides for the practical applicability of such a logic.

There is also a fatal error plaguing the intentionalist approach in this area, and it is recognized by John Dewey in his *Human Nature and Conduct*. Dewey argues that one must not suppose that preference emerges only after deliberation has taken place, or i.e., after the stage of the "weighing of reasons," as von Wright would have it. Rather, Dewey perceives choice as the result of choosing among many possible alternative modes of action as already perceived preferences. Thus, Dewey's position undercuts von Wright's operational distinction between choice and preference. For he is saying that preference is not the result of something *different*, like deliberation upon reasons, *ceteris paribus*, being manifested as the ground of choice which results in preference. Rather, choice is the manifestation of the strength of one preference predominating over another.[8] Generally, the virtue of Dewey's analysis is that it

---

[8] Dewey, John, *Human Nature and Conduct*, Henry Holt and Company, New York, 1922, p. 194.

neutralizes the traditional subject/object split found in theories of experience, and thus redefines experience as a resultant and relevant symbiosis between organism and environment. Thus, with reference to his account of preference, one sees that by explaining choice in terms of competing preferences, Dewey is providing an alternative which escapes allusions to context-free mental states, suggested in expressions such as the "weighing of reasons," under conditions of *ceteris paribus*.

In an important passage of *Human Nature and Conduct* Dewey is found to say as follows:(italics added)[9]

> "... Choice is made as soon as some habit, or some combination of elements of habits and impulse, finds a way fully open. Then energy is released. The mind is made up, composed, unified. As long as deliberation pictures shoals or rocks or troublesome gales as marking the route of a contemplated voyage, deliberation goes on. But when the various factors in action fit harmoniously together, when imagination finds no annoying hindrance, when there is a picture of open seas, filled sails and favoring winds, the voyage is definitely entered upon. This decisive direction of action constitutes choice. *It is a great error to suppose that we have no preferences until there is choice.* We are always biased beings, tending in one direction rather than another. The occasion of deliberation is an *excess* of preferences, not natural apathy or an absence of likings. We want things that are incompatible with one another; therefore we have to make a choice of what we *really* want, of the course of action, that is, which most fully releases activities. *Choice is not the emergence of preference out of indifference. It is the emergence of a unified preference out of competing preferences.* ..."

More strikingly, Dewey proceeds to invert the role of reason with respect to imaginative activity. This is surely contrary to the spirit of von Wright's position, where the latter argues for the view that there is an attitude of preference which is antecedent and separate from choice, and that this stage of the "weighing of reasons" is the object of the intentionalist analysis of the formal properties of the preference-relation By contrast, Dewey places the role of ratiocination at the end of successful action, observing that the proper role of reason is to synthesize different modes of action into a consistent pattern, hav-

---

[9] ibid., pp. 192-193.

ing some internal coherence. Again, in a telling passage Dewey's words hit home with a disarming certitude: (italics added)[10]

> "... The elaborate systems of science are born not of reason but of impulses at first slight and flickering; impulses to handle, move about, to hunt, to uncover, to mix things separated and divide things combined; to talk and to listen. Method is their effectual organization into continuous dispositions of inquiry, development and testing. *It occurs after these acts and because of these consequences. Reason, the rational attitude, is the resulting disposition, not a ready-made antecedent which can be invoked at will and set into movement.* The man who would intelligently cultivate intelligence will widen, not narrow, his life of strong impulses while aiming at their happy coincidence in operation."

This entire inquiry indicates that studies in the formalization of preferences have a profound role to play in the pursuit of theories of value. However, by-in-large, such inquiries have been neglected or given little heed in the mainstream of philosophical inquiry. A result has been that the pursuit of a viable utilitarian theory of ethics has suffered. It would appear that an approach such as is sketched by R.M. Martin, with emphasis on extensionalized pragmatics, coupled with a logic of events, can do much to clarify the sense of value discourse in this area. Moreover, this kind of study should enable future analysts to pursue investigations into the nature of preference as an activity which resides totally and solely in the domain of philosophy.

The ramification of this direction of research is virtually unlimited in the variety of its applicability. For example, one finds its use in theories of interpretation. This is a point Donald Davidson has recently noticed, in that a theory of interpretation requires a theory of preference at its basis.[11] Here an understanding of the coherence of interpretation comes into play as the fabric which an analysis of preference spins off. Along similar lines, one sees the possibilities of applying inquiries into the formal character of preference expressions to studies concerning the nature of experimental procedures in scientific testing, the priorities one should adopt in issues dealing with environmental ethics, or medical ethics, etc., etc.

---

[10] ibid., p. 196.
[11] Davidson, Donald, *Inquiries Into Truth and Interpretation*, Clarendon Press, Oxford, 1984, p. 148.

Selected Bibliography

Apostel, Leo, "Logique Inductive, Modalities Epistemiques et Logique de la Preference," *Rev. Int. Phil.* 25, (1971), 78-100.

Aqvist, Lennart, "Deontic Logic Based on a Logic of 'Better'," *Acta Philosophica Fennica*, fasc. XVI, (Helsinki, 1963), pp.285-290.

Aqvist, Lennart, "A Binary Primitive in Deontic Logic," *Logique et Analyse*, 19, (1962), pp. 90-97.

Aristotle, (4th cent. B.C.), *Topics*, book iii.

Arrow, K.J., *Social Choice and Individual Values*, New York, 1951.

Berka, K., "On Logical and Methodological Problems in the Theory of Preference (In Czech)," *Teor Metod*, 8, (1976), 45-68.

Bieltz, Petre, "Logical Foundations of Social Decision," *Philosophie et Logique*, 21, (1977), 265-273.

Bohnert, Herbert G., "The Logical Structure of the Utility Concept," in *Decision Processes*, edited by R.M. Thall, C.H. Coombs, and R L. Davis, John Wiley and Sons, Inc , New York, 1954.

Bolle, Friedel, "On Sen's Second-Order Preferences, Morals and Decision Theory," *Erkenntnis*, 20, (1983), 195-206.

Bolker, Ethan, "A Simultaneous Axiomatization of Utility and Subjective Probability," *Philosophy of Science*, 34, (1967), 333-340.

Brentano, Franz, *The Origin of Our Knowledge of Right and Wrong*, Routledge and Kegan Paul, New York, 1969.

Brogan, A P., "The Fundamental Value Universal," *Journal of Philosophy, Psychology and Scientific Methods*, vol. 16, (1919).

Burros, Raymond H., "Complementary Relations in the Theory of Preference," *Theory and Decision*, 7, (1976), 181-190.

Castaneda, Hector Neri, "On the Logic of 'Better'", review, *Philosophy and Phenomenological Research*, 1958, Vol. 19.

Chisholm, Roderick M., *Realism and the Background of Phenomenology*, The Free Press, New York, 1960.

Chisholm, Roderick M., "Epistemic Reasoning and the Logic of Epistemic Concepts," in *Logic and Philosophy*, von Wright, G.H., The Hague: Nijhoff, 71-78.

Chisholm, Roderick M, Keim, Robert G., "A System of Epistemic Logic," *Ratio*, 14, (1972), 99-115.

Chisholm, Roderick M., "On A Principle of Epistemic Preferability," *Philosophy and Phenomenological Research*, 30, (1969), 294-301.

Craven, John, "Liberalism and Individual Preferences," *Theory and Decision*, 14, (1982), 351-360.

Cresswell, M.J., "A Semantics for a Logic of 'Better'," *Logical Analysis*, 14, (1971), 777-782.

Dalen, Dirk Van, "Variants of Rescher's Semantics for Preference Logic and Some Completeness Theorems," *Studia Logica*, 33, (1974), 163-181.

Danielsson, Sven, "Preference and Obligation," *Studies in the Logic of Ethics*, Uppsala: Filos Foreningen, 1968.

Davidson, Donald; McKinsey, J.C.C. Suppes, Patrick, "Outlines of a Formal Theory of Value," *Philosophy of Science*, 22, (1955), 140-160.

Davidson, Donald, Suppes, Patrick, *Decision Making — An Experimental Approach*, Stanford University Press, Stanford, California, 1957.

Davidson, Donald, *Inquiries Into Truth and Interpretation*, Clarendon Press, Oxford, 1984.

Di Bernardo, G., "Thetic and Prohairetic Normative Systems," *Poznan Studies*, 5, (1979), 87-118.

Domotor, Zoltan, "Axiomatization of Jeffrey Utilities," *Synthese*, 39, (1978), 165-210.

Farquhar, Peter H., "Advances in Multiattribute Utility Theory," *Theory and Decision*, 12, (1980), 381-394.

Fishburn, Peter C., *The Foundations of Expected Utility*, Boston, Reidel, 1982.

Fishburn, Peter C. "A Theory of Subjected Utility with Vague Preferences," *Theory and Decision*, 6, (1975), 287-310.

Franke, Gunter, "Expected Utility with Ambiguous Probabilities and 'Irrational' Parameters," *Theory and Decision*, 9, (1978), 267-283.

Friedland, Edward I., Cimbala, Stephen J., "Process and Paradox: The Significance of Arrow's Theorem," *Theory and Decision*, 4, (1973), 51-62.

Gardenfors, Peter, "Some Basic Theorems of Qualitative Probability," *Studise Logica*, 34, (1975), 257-264.

Hallden, Soren, *On the Logic of 'Better'*, Uppsala, 1957, "Preference Logic and Theory Choice," *Synthese*, 16, (1966), 307-320.

Hallden, Soren, *Logik Ratt och Moral*, 1969, Filosofiska Studier tillagnade, Manfred Moritz, 1969.

Hallden, Soren, *The Foundation of Decision Logic*, Library of Theoria, No.14, Lund, 1980.

Hansson, Bengt, "Choice Structures and Preference Relations," *Synthese*, 18, (1968), 443-458.

Hansson, Bengt, "Fundamental Axioms for Preference Relations," *Synthese*, 18, (1968), 423-42.

Harsanyi, John C , "Rule Utilitarianism and Decision Theory," *Erkenntnis*, 11, (1977), 25-53.

Hochberg, Gary M., "Extrinsic Epistemic Preferability," *Philosophical Studies*, 23, (1972), 76-83.

Hooker, Michael, "Chisholm's Theory of Knowledge," *Philosophia* (Israel), 7, (1978) 489-500.

Houthakker, H.S., "The Logic of Preference and Choice," in A.T. Tymieniecka (ed.), *Contributions to Logic and Methodology in Honor of J.M. Bochenski* (Amsterdam, 1965), pp. 193-207. (An attempt to draw together the interests of logicians and economists.)

Houthakker, H.S., "Revealed Preference and the Utility Function," *Economica*, vol. 17, (1950).

Huber, Oswald, "Nontransitive Multidimensional Preferences: Theoretical Analysis of a Model," *Theory and Decision*, 10, (1979), 147-165.

Iwanus, Boguslaw, "Concerning the So Called System BD of Aqvist," *Studies of Logic*, 33 (1974), 339-343.

Jeffrey, Richard C., "Preference Among Preferences," *Journal of Philosophy*, 71, (1974), 377-391

Jeffrey, Richard C., *The Logic of Decision*, (First Edition) New York: McGraw-Hill, 1965; (Second Edition), University of Chicago Press, 1983.

Jeffrey, Richard C., "Ethics and the Logic of Decision," *Journal of Philosophy*, (1965), 528-538.

Jennings, R.E., "'Or,'" *Analysis*, 26, (1966), 181-184.

Jennings, R.E., "Preference and Choice as Logical Correlates," *Mind*, 76, (1967), 556-557.

Katkov, Georg, *Untersuchungen zur Werttheorie und Theodizee*, Brunn, 1937.

Kneale, William, *Probability and Induction*, Oxford University Press, 1963.

Kraus, Oskar, *Die Werttheorien*, Brunn, 1937.

Kron, Aleksandar, Milovanovic, Veselin, "Preference and Choice," *Theory and Decision*, 6, (1975), 185-196.

Kyburg, Henry E. Jr., Smokler, Howard E., *Studies in Subjective Probability*, John Wiley & Sons Inc., 1964.

Marshall, Alfred, *Principles of Economics*, 9th edition, New York, The Macmillan Company, 1961, (Chapter IIIa).

Martin, R.M., "Toward a Logic of Intensions," *Synthese*, 15, (1963), 81-102.

Martin, Richard M., *Intension and Decision*, Englewood Cliffs, N.J., Prentice-Hall, 1963.

Martin, Richard M., "On Pragmatics, the Meta-Theory of Science, and Subjective Intensions," in *Logic and Art, Essays in Honor of Nelson Goodman*, edited by Richard Rudner and Israel Scheffler, Bobbs-Merrill Company, Inc., 1972.

Moore, G.E., "A Reply to My Critics." in P.A. Schlipp (ed.), *The Philosophy of G.E. Moore*, Evanston, 1942.

Moutafakis, Nicholas J., "Rescher's Logic of Preference and Linguistic Analysis," *Logique et Analyse*, 25, (1982), 135-165.

Mullen, John D., "Does the Logic of Preference Rest on a Mistake?" *Metaphilosophy*, 10, (1979), 247-255.

Nowak, L., "Dobro I Zlo W Swietle Logiki Preferencji." *Etyka*, 12, (1973), 157-161.

Packard, Dennis, "Plausibility Orderings and Social Choice" *Synthese*, 49, (1981), 415-418.

Packard, Dennis J., "Cyclical Preference Logic," *Theory and Decision*, 14, (1982), 415-426.

Parks, R.Z., "On Jennings on Von Wright on Preference," *Mind*, 80, (1971), 288-289.

Pattanaik, Prasanta K., "Collective Rationality and Strategy-Proofness of Group Decision Rules," *Theory and Decision*, 7, (1976), 191-203.

Peklo, B.T., "A Note on Logical Values in Classical Logic and Logic of Preference," *Teor Metod*, 5, (1973), 151-153.

Ramsey, Frank P., *The Foundations of Mathematics and Other Logical Essays*, London, Kegan Paul, 1931.

Rescher, N., "Choice Without Preference," *Kantstudien*, 51, 142-175.

Rescher, Nicholas, *Introduction to Value Theory*, Englewood Cliffs, N.J., Prentice-Hall, 1969.

Rescher, Nicholas, *The Logic of Decision and Action*, Pittsburgh: University of Pittsburgh Press, 1968.

Saito, Setsuo, "Modality and Preference Relation," *Notre Dame Journal of Formal Logic*, 14, (1973), 387-391.

Scheler, Max, *Der Formalismus in der Ethik und die materiale Wertethik*, Halle, 1921.

Schumm, George F., "Remark on a Logic of Preference," *Notre Dame Journal of Formal Logic*, 16, (1975), 509-510.

Scwartz, Herman, *Psychologie des Willens zur Grundelegung der Ethik*, Leipzig, 1900.

Seeskin, Kenneth R., "Many-Valued Logic and Future Contingencies," *Logique et Analyse*, 14, (1971), 759-773.

Seidenfeld, Teddy, Schervish, Mark, "A Conflict Between Finite Additivity and Avoiding Dutch Book," *Philosophy of Science*, 50, (1983), 398-412.

Smith, Tony E., "Rationality of Indecisive Choice Functions on Triadic Choice Domains," *Theory and Decision*, 10, (1979), 113-129.

Sorensen, Roy A., "Is Epistemic Preferability Transitive?," *Analysis*, 41, (1981), 122-123.

Tversky, Amos, "A Critique of Expected Utility Theory: Descriptive and Normative Considerations," *Erkenntnis*, 9, (1975), 163-173.

Uslaner, Eric M., "The Paradox of Voting with Indifference," *Philosophica*, 20, (1977), 7-33.

Von Kutschera, Franz, "Semantic Analyses of Normative Concepts," *Erkenntnis*, 9, (1975), 195-218.

Von Wright, Georg Henrik, "Remarks on the Epistemology of Subjective Probability," in *Logic, Methodology and Philosophy of Science*, Proceedings of the 1960 International Congress, (Stanford, 1962).

Von Wright, Georg Henrik, *The Logic of Preference: An Essay*, Edinburgh: Edinburgh University Press, (1963).

Von Wright, G.H., "The Logic of Preference Reconsidered," *Theory and Decision*, 3, (1972), 140-167.

Walhout, Donald, "The Logic of Aesthetic Judgments," *Midwest Journal of Philosophy*, Spring 1973, 43-52.

White, Alan R., *Truth*, Anchor Books, 1970.

White, Alan R., *Modal Thinking*, Cornell University Press, Ithaca, New York, 1975.

Willing, Anthony, "A Note on Rescher's 'Semantic Foundations for the Logic of Preference'," *Theory and Decision*, 7, (1976), 221-229.

Wiseman, Charles, "The Theory of Modal Groups," *Journal of Philosophy*, 67, (1970), 367-376.

Zadeh, L.A., "Linguistic Characterization of Preference Relations as a Basis for Choice in Social Systems," *Erkenntnis*, 11, (1977), 383-410.

Name Index

A

Ackerman, Robert 96.

Aqvist, Lennart 4.-5., 191., 192., 193.-198., 199., 218., 229.

Aristotle, 1., 57.-76., 94.-95., 196., 202.

Arrow, Kenneth, 6., 7., 28.-29., 183.

B

Bayes, Thomas 17.-18.

Bohnert, Herbert 3., 39.-40., 42.-54., 58., 212., 222., 237., 255.

Bolker, Ethan 143.-144., 151.

Boole, George 228.-229.

Brentano, Franz 6., 135., 192., 202.-204., 275.

Brogan, A.P. 199., 200.

Brutus, 229.-230.

C

Caesar, 229.-230.

Carnap, Rudolf 3., 43., 45., 47.-51., 52., 53.

Castaneda, Hector Neri 178., 179.-180., 181., 189.

Chisholm, Roderick 5., 7., 89., 111., 178., 180., 181., 191., 192., 198.-209., 210., 227., 250-252., 275.

Chomsky, Noam 212.-213.

Chopin, Frederic F. 96.

Columbus, Christopher 126.-127.

D

Davidson, Donald 1-2, 3., 13.-15., 19., 29.-42., 47.-49., 53., 59., 128.-129., 131.-133., 168.-169., 174., 207., 210., 212., 229., 230., 237., 238.-239., 255., 269., 272., 275., 278.

Davis, Robert L. 53.

DeFinetti, Bruno 6., 135.

Dewey, John, 8., 276.-278.

E

Evans, J.D.G., 73.

F

Fishburn, Peter C., 9.

G

Geach, Peter 133.-134.

Goodman, Nelson 111.

H

Hallden, Soren 4., 89., 167.-190., 191., 197., 198., 199., 207., 209., 210., 211., 217., 255., 274., 276.

Hamlyn, D.W. 75.-76.

Hansson, Begnt, 5., 191., 192., 209.-219., 271.

Hochberg, Gary 6., 257.-268.

Hume, David 13.

J

Jeffrey, Richard C., 2., 4., 7., 17., 119.-164., 178., 180.-181., 207., 211., 254.

Jennings, R.E. 224.n., 236.

K

Keynes, John M. 1., 16.-17., 26.

Kneale, William 45.-47.

Kyburg, Henry 135.-136.

L

Lewis, C.I., 115., 119., 156.

M

Marshall, Alfred 39., 97.-98.

Name Index

Martin, Richard M., 3., 7., 44., 52., 57.-88., 89., 122., 124., 129., 207., 210., 222., 223., 237., 268., 269., 271., 273.-275., 278.
Meinong, Alexis 6., 7., 192., 202., 204.-205., 227.
Montague, Richard, 86.-88., 273.-274.
Moore, George E. 93.-94., 172., 176., 199.
Mullen, John D., 8., 269.-272.

O
Orozco, 87.

P
Parks, R.Z. 224.n.
Peirce, Charles S. 20., 27., 260.
Plato, 1., 62.-63.
Pythagoras 101.

Q
Quine, Willard van Orman 102.

R
Ramsey, Frank P., 1., 3., 5., 6., 13.-29., 30., 53., 54., 123.-124., 174., 199., 207., 210., 237., 255., 269., 272., 275.
Rescher, Nicholas 2., 3., 33., 57.-58., 60., 74., 89.-117., 133., 134., 188., 210., 212., 252., 254., 275.
Reichenbach, Hans 35.
Russell, Bertrand 48.

S
Schick, Frederick 122.-123.
Scheler, Max 199.
Schlick, Moritz 172.
Schwartz, Herman 199.
Sosa, Ernest 5., 7., 89., 178., 180., 181., 191., 192., 198-209., 210., 250.-252., 275.
Stalnaker, Robert 111.-115., 117.
Stevenson, Charles L. 172.
Suppes, Patrick, 13., 19., 29.-42., 174.

T
Titian, 87.

V
Vickers, 17.-18., 86.-88.
Von Neumann, 7., 16., 37., 39., 42., 43.-44., 48., 51., 53., 87.
Von Wright, Georg Henrik, 1., 2., 5.-6., 7., 16., 23.-27., 54., 81.-84., 89., 116.-117., 135., 167.-168., 191., 193., 194., 206., 207., 209., 210., 211., 221.-255., 269., 270.-276.

W
Weil, E., 61., 76.
White, Alan 89.-90., 93.-94., 109.-110., 133., 228., 232.
Willing, Anthony 102.-108.
Wittgenstein, Ludwig 15., 27., 100.-101., 162.

Subject Index

A

Axiomatization 13.-14., 21.-22., 24., 29.-32., 37.-40., 42., 58., 63., 74., 89., 108., 116.-117., 122., 138.-149., 177., 179., 181., 183., 206.-210., 211., 213.-215., 217.-218., 232., 239., 257.

C

Certeris paribus 4., 6., 82., 194., 225., 227., 232., 237., 238., 241., 242., 244., 254., 255., 276-277.

Choice 3., 4., 7.-8., 13., 15., 18., 19., 21., 23., 26., 28., 31.-33., 34., 38., 44., 45., 47., 55., 58., 59.-60., 63., 64., 65., 69., 70., 72., 73., 77., 78., 82., 83.-84., 86., 116., 117., 120., 127.-129., 137., 148., 149., 153., 155., 159., 163., 167., 172., 203.-204., 223., 224., 225., 227., 236., 268., 269., 270., 271., 272., 273., 274., 275., 276., 277.

D

Degree of belief 16.-22., 25., 87., 120.
De dicto / De re 5.-6., 83., 232., 235.
Deontic Logic 4., 171., 172., 191., 192., 193., 194., 198., 241.

E

Extensionality 3.-6., 7., 8.-9., 13., 18., 21., 27.-28., 38., 44., 47., 48., 51.-53., 54.-55., 58.-59., 62.-63., 66., 67.-68., 76., 79., 81.-82., 83., 85.-86., 88., 92., 99., 101., 109.-110., 119., 124., 129., 134., 137., 156.-157., 164., 183., 214.-215., 223., 225.-226., 232., 237., 241., 243., 247., 255., 260., 266.-268., 269., 271., 273.-276., 278.

F

Frequency Probability 44.-50.

I

Intensions 89., 257., 259.
Intentionality 4.,5., 9., 81., 82., 110., 111., 115., 116., 127., 129., 156., 164., 176., 230., 237., 258., 275., 276., 277.

M

Mathematics 28., 31., 91., 101.-102., 105.-107., 151., 188.
Measurement 29.-33.
Metalanguage 33.

P

Probability 1., 3., 4., 6., 7., 14., 16., 17.-18., 19., 21., 25., 26., 27., 32., 37., 39., 40., 41., 42., 43., 44.-51., 53.-54., 58., 98., 120., 121., 122.-124., 134.-135., 136., 137.-138., 139.-140., 141., 145., 146., 147.-148., 150.-151., 154., 159.-160., 164., 169., 235.-237., 255.

Propositions 4., 5., 6., 15., 16., 17., 18., 20., 21., 22., 23., 24., 26., 43., 44., 47., 48., 49., 50., 51., 52., 53., 58., 61., 82., 83., 84., 85., 92., 99.-100., 101., 103., 109.-110., 119., 120., 122., 123., 126., 127., 129.-130., 131.-133., 136., 138., 139., 140., 142., 143., 146.-147., 148., 149., 150.-151., 152., 153., 154., 155., 164., 170., 173., 175., 178., 179., 182.-183., 184., 197.-198., 199., 203., 204., 206., 208., 210., 211.-212., 221., 223., 226., 228., 229., 230., 231.-232., 234., 241., 242., 243., 246., 250., 250., 260., 262.-263.-264., 265.-266.

S

States of Affairs 5., 6., 17., 22., 26.-27., 35., 43., 60., 86., 90., 95., 98.-99., 100., 102.-103., 106., 107., 135., 136.-137., 138., 155.-156., 160., 168.-169., 176., 178., 179.-180., 181.-183, 186., 192., 199., 201., 205., 206., 223., 225., 226., 230.-232., 233., 234., 238., 240., 241., 251., 252., 258., 260.-261., 262.-263., 264.

T

Transitivity 28.

MIX
Papier aus verantwortungsvollen Quellen
Paper from responsible sources
FSC® C105338

If you have any concerns about our products,
you can contact us on
ProductSafety@springernature.com

In case Publisher is established outside the EU,
the EU authorized representative is:
**Springer Nature Customer Service Center GmbH
Europaplatz 3, 69115 Heidelberg, Germany**

Printed by Libri Plureos GmbH
in Hamburg, Germany